Universitext

M. Schreiber

Differential Forms
A Heuristic Introduction

Springer-Verlag
New York Heidelberg Berlin

M. Schreiber

Department of Mathematics
The Rockefeller University
1230 York Ave.
New York, N.Y. 10021

AMS Subject Classifications: 26-01, 26A57, 26A60, 26A66

Library of Congress Cataloging in Publication Data

Schreiber, Morris, 1926–
  Differential forms.

  (Universitext)
  Bibliography: p.
  1. Differential forms. I. Title.
QA381.S4   515′.37   77–14392

9 8 7 6 5 4 3 2 1

ISBN-13: 978-0-387-90287-6   e-ISBN-13: 978-1-4612-9940-0
DOI: 10.1007/978-1-4612-9940-0

Preface

A working knowledge of differential forms so strongly illuminates the calculus and its developments that it ought not be too long delayed in the curriculum. On the other hand, the systematic treatment of differential forms requires an apparatus of topology and algebra which is heavy for beginning undergraduates. Several texts on advanced calculus using differential forms have appeared in recent years. We may cite as representative of the variety of approaches the books of Fleming [2], [1] Nickerson-Spencer-Steenrod [3], and Spivak [6]. Despite their accommodation to the innocence of their readers, these texts cannot lighten the burden of apparatus exactly because they offer a more or less full measure of the truth at some level of generality in a formally precise exposition. There is consequently a gap between texts of this type and the traditional advanced calculus. Recently, on the occasion of offering a beginning course of advanced calculus, we undertook the experiment of attempting to present the technique of differential forms with minimal apparatus and very few prerequisites. These notes are the result of that experiment.

Our exposition is intended to be heuristic and concrete. Roughly speaking, we take a differential form to be a multi-dimensional integrand, such a thing being subject to rules making change-of-variable calculations automatic. The domains of integration (manifolds) are explicitly given "surfaces" in Euclidean space. The differentiation of forms (exterior

---

(1) Numbers in brackets refer to the Bibliography at the end.

differentiation) is the obvious extension of the differential of functions, and this completes the apparatus. To avoid the geometric and not quite elementary subtleties of a correct proof of the general Stokes formula we offer instead a short plausibility argument which we hope will be found attractive as well as convincing. This is one of several abbreviations we have made in the interests of maintaining an elementary level of exposition.

The prerequisite for this text is a standard first course of calculus and a bit more. The latter, though not very specific, may be described as some familiarity with Euclidean space of k dimensions, with k-by-k matrices and the row-by-column rule for multiplying them, and with the simpler facts about k-by-k determinants. Serious beginning undergraduates seem generally to possess this equipment at the present time. Linear algebra proper is not required, except at one place in Chapter 6, where we must diagonalize a real symmetric matrix. For this theorem, and for several other facts of algebra (such as those mentioned above), we offer references to the text [5] of Schreier and Sperner. There the matters in question are well presented without prerequisites. For analytical matters we provide citations to Courant [1]. We have tried to design the text so that, with the books of Courant and Schreier-Sperner as his only other equipment, the industrious reader working alone will find here an essentially self-contained course of study. However, the better use of this text is probably its obvious one as part of a modern sophomore or junior course of advanced calculus.

The content of each Chapter is clear from the Table of Contents, with two exceptions: in 6.2 we give the theorem on the geometric and arithmetic means, and in 7.3 we prove the isoperimetric inequality. Our notations, all standard, are listed on page (x). The symbol n.m(k) means Formula (k) in Section n.m.

It is a pleasure to acknowledge several debts of gratitude: to P.A. Griffiths, who encouraged the project and suggested the inclusion of "something on integral geometry"; to Mary Ellen O'Brien, who gave the manuscript its format in the course of typing it; and to the students, who were willing to participate in an experiment.

M. Schreiber
21 July 1977

Table of contents

Notations

$a \in A$	a is a member of the set A
$\vec{x}$	a vector
$\|\vec{x}\|$	the length of $\vec{x}$
$\vec{x} \cdot \vec{y}$	the scalar product of $\vec{x}$ and $\vec{y}$
$\vec{x} \times \vec{y}$	the vector product of $\vec{x}$ and $\vec{y}$
$\binom{n}{k}$	the binomial coefficient $\binom{n}{k} = \dfrac{n!}{(n-k)!\,k!}$
$\mathbb{R}$	the set of real numbers
$\mathbb{R}^k$	Euclidean space of dimension k
$f : \mathbb{R}^n \rightarrow \mathbb{R}^m$	m-vector-valued function of an n-vector argument
$Z(\phi)$	the set of zeros of $\phi : \mathbb{R}^k \rightarrow \mathbb{R}^1$
$f \circ g$	the composition $f \circ g(x) = (f(g(x))$ of functions f and g
$\omega \wedge \tau$	the wedge product of differential forms ω and τ
Λ^r	the space of r-forms
$\mathbb{W}$	the set of vector fields
$\mathbb{S}$	the set of scalar fields
$|T|$	the determinant of the matrix T
${}^t T$	the transpose of the matrix T
$\mathrm{tr}\, T$	the trace of the matrix T

Chapter 1
Partial differentiation

1.1 Partial Derivatives

We denote by $\mathbb{R}^k$ the set of real ordered k-tuples
$\vec{x} = (x_1, x_2, \ldots, x_k)$.   Such a k-tuple is called a k-vector, the
numbers $x_1, x_2, \ldots$ being its components.   k-vectors are added and
multiplied by scalars in the component-wise function familiar in the
plane and in three-space.   In this notation the plane and three-space
are denoted $\mathbb{R}^2$ and $\mathbb{R}^3$ respectively.

The inner (or scalar) product $\vec{x} \cdot \vec{y} = \sum_i x_i y_i$ of k-vectors $\vec{x}$
and $\vec{y}$ determines length and angle in $\mathbb{R}^k$ as follows. The length
$\|\vec{x}\|$ of $\vec{x}$ is $\|\vec{x}\| = \sqrt{\vec{x} \cdot \vec{x}}$, and the angle θ between $\vec{x}$ and $\vec{y}$
is defined by the relation (law of Cosines) $\text{Cos } \theta = \vec{x} \cdot \vec{y}/\|\vec{x}\| \cdot \|\vec{y}\|$.
These are exact analogs for $\mathbb{R}^k$ of the corresponding constructions
in $\mathbb{R}^2$ and $\mathbb{R}^3$. In particular, $\vec{x} \in \mathbb{R}^k$ is called a unit vector
if $\|\vec{x}\| = 1$, and k-vectors $\vec{x}, \vec{y}$ are orthogonal if $\vec{x} \cdot \vec{y} = 0$.

A function of n variables may be viewed as a function of an
n-vector argument or variable. We shall be concerned also with functions
taking vector values. The notation $f : \mathbb{R}^n \to \mathbb{R}^m$ signifies that f
is a function of an n-vector argument taking m-vector values. Since

a 1-vector is just a number, a function $f : \mathbb{R}^n \to \mathbb{R}^1$ is a scalar-valued function of n variables.

Take as orthogonal reference frame in $\mathbb{R}^k$ the unit vectors $\vec{e}_1 = (1,0,\ldots,0)$, $\vec{e}_2 = (0,1,0,\ldots,0)$, $\ldots$, $\vec{e}_k = (0,\ldots,0,1)$. The lines on which they lie are then a system of axes for a Cartesian coordinate system in $\mathbb{R}^k$. Every point $\vec{x} \in \mathbb{R}^k$ has a unique expression $\vec{x} = \sum x_i \vec{e}_i$ in this reference frame, its Cartesian coordinates in this frame being $(x_1, x_2, \ldots, x_k)$. To each coordinate direction is associated a partial differential operator $\dfrac{\partial}{\partial x_i}$, which acts upon scalar-valued functions $f : \mathbb{R}^k \to \mathbb{R}^1$ of a vector argument thus:

$$[\frac{\partial}{\partial x_i} f](\vec{x}) = \lim_{\delta \to 0} \frac{f(\vec{x} + \delta \vec{e}_i) - f(\vec{x})}{\delta} . \tag{1}$$

Note well that the result of applying the operator $\dfrac{\partial}{\partial x_i}$ to a function $f : \mathbb{R}^k \to \mathbb{R}^1$ is another function $[\frac{\partial}{\partial x_i} f] : \mathbb{R}^k \to \mathbb{R}^1$ of the same type. We will denote this new function, whenever possible, by the short notation f_i. Thus

$$f_i(\vec{x}) = [\frac{\partial}{\partial x_i} f](x) ; \tag{2}$$

and f_i is called the ith partial derivative of f. Its geometric significance is as follows. Given $f : \mathbb{R}^k \to \mathbb{R}^1$, we may interpret the equation

$$z = f(\vec{x}) \tag{3}$$

as defining a k-dimensional surface in $\mathbb{R}^{k+1}$. If we fix all but the ith coordinate of a point $\vec{x} \in \mathbb{R}^k$, and let the ith coordinate x_i vary freely, we generate a straight line in $\mathbb{R}^k$ passing through x and lying parallel to $\vec{e}_i$. Its coordinates are $(x_1, x_2, \ldots, x_{i-1}, t, x_{i+1}, \ldots, x_k)$, with $-\infty < t < +\infty$. Its image under f is a 1-dimensional curve (it has one degree of freedom; namely, the variation of t) lying on the surface (3). The slope of this curve as a function of t is

$$f_i(x_1, x_2, \ldots, x_{i-1}, t, x_{i+1}, \ldots, x_k) .$$

Since the partial derivatives f_i of a function $f: \mathbb{R}^k \to \mathbb{R}^1$ are again functions of the same type, they too may be differentiated partially. The result of differentiating f_i by its jth argument may (and will, whenever possible) be denoted f_{ij}. The concordance with the standard notation is

$$f_{ij} = \frac{\partial^2}{\partial x_j \partial x_i} f = \frac{\partial}{\partial x_j}\left(\frac{\partial}{\partial x_i} f \right) . \tag{4}$$

One should note the reversal of order of the subscripts. Since the f_{ij} are again functions of the same type, they may be differentiated partially; denote these functions by f_{ijk}. In this hierarchy, the f_i are called first partials, the f_{ij} second partials, and so on. We are assuming for the purposes of this discussion that all limits involved (they are all of the general form (1)) exist. This being assumed, there are in principle k first partials $f_1, f_2, \ldots, f_k$; k^2 second partials $f_{11}, f_{12}, \ldots, f_{1k}, f_{21}, f_{22}, \ldots, \ldots, f_{kk}$; k^3 third partials; and so on.

On the other hand, it is not hard to show, using the mean value theorem,
that when all partial derivatives involved are themselves continuous
functions of their several variables, then the order in which the
differentiations are done does not matter.[1] For example, $f_{12} = f_{21}$
if both are continuous functions; similarly $f_{112} = f_{121} = f_{211}$ if
all are continuous. Thus the number of distinct higher partials is
sharply reduced if they are continuous. The rule for the equality of
mixed higher partials may be stated thus: two higher partials are equal
(when they are continuous functions) if they involve the same indices
with the same multiplicities.

1.2 Differentiability, Chain Rule

By definition a real function $f : \mathbb{R}^1 \to \mathbb{R}^1$ of a real argument is
differentiable at x if

$$\frac{f(x+h) - f(x)}{h} = f'(x) + \varepsilon(h) , \qquad (1)$$

$$\varepsilon(h) \to 0 \quad \text{as} \quad h \to 0 ; \qquad (2)$$

which is to say, the increment

$$(\Delta f)(x,h) = f(x+h) - f(x)$$

(1) See Courant [1], volume II, pp. 55-58.

is approximated by the differential

$$(df)(x,h) = f'(x) \cdot h \tag{3}$$

so well that the error $\eta(h) = h \cdot \varepsilon(h)$ vanishes faster than h, $\lim_{h \to 0} \frac{\eta(h)}{h} = 0$. Note that the differential is a linear function of the increment h.

Suppose, for a given function $f : \mathbb{R}^1 \to \mathbb{R}^1$ and a given $x \in \mathbb{R}^1$, that there exists a constant A and a function $\alpha : \mathbb{R}^1 \to \mathbb{R}^1$ such that

$$(\Delta f)(x,h) = Ah + \alpha(h),$$

$$\tag{4}$$

$$\lim_{h \to 0} \frac{\alpha(h)}{h} = 0 ;$$

which is to say, the increment $(\Delta f)(x,h)$ can be approximated with the stated accuracy by a linear function of h. This would imply at once that f is differentiable at x and that $f'(x) = A$.

Putting together the two foregoing paragraphs, we see that, for functions of one variable, differentiability is the same as linear approximability. The generalization of differentiability to functions of several variables is made by generalizing to several variables the idea of linear approximability, as follows.

A function $f : \mathbb{R}^k \to \mathbb{R}^1$ is (by definition) differentiable at $x \in \mathbb{R}^k$ if there exist constants $A_1, A_2, \ldots, A_k$ and a function $\alpha : \mathbb{R}^k \to \mathbb{R}^1$ such that

$$(\Delta f)(\vec{x},\vec{h}) = \sum_i A_i h_i + \alpha(\vec{h}) , \qquad (5)$$

$$\lim_{\vec{h} \to \vec{0}} \frac{\alpha(\vec{h})}{\|\vec{h}\|} = 0 , \qquad (6)$$

where $\vec{h} = (h_1, h_2, \ldots, h_k)$ is the increment vector and $(\Delta f)(\vec{x},\vec{h}) = f(\vec{x}+\vec{h}) - f(\vec{x})$ is the corresponding increment of f.

We have replaced the linear function Ah of (4) by a linear function of the components $h_1, \ldots, h_k$ of the increment vector $\vec{h}$, and "$h \to 0$" becomes "$\vec{h} \to \vec{0}$". This is the exact analog of (4). Note that for any $\vec{x} \in \mathbb{R}^k$ one has

$$|x_j| \leq \|\vec{x}\| \leq \sqrt{k} \cdot \text{Max}\{|x_1|, \ldots, |x_k|\}, \qquad (7)$$

for $x_j^2 \leq \sum x_i^2 \leq k \cdot \text{Max}\{x_1^2, \ldots, x_k^2\}$. Therefore all components of $\vec{x}$ are small independently if and only if $\|\vec{x}\|$ is small, and so $\vec{h} \to \vec{0}$ if and only if $\|\vec{h}\| \to 0$ if and only if $h_j \to 0$ for each j.

Assume f is differentiable in this sense, and put $\vec{h} = \delta \vec{e}_i$ in (5). This yields

$$\frac{f(\vec{x}+\delta\vec{e}_i) - f(\vec{x})}{\delta} = A_i + \frac{\alpha(\vec{h})}{\delta} ,$$

whence the limit as $\delta \to 0$ (i.e., $\vec{h} \to \vec{0}$) exists, and $A_i = f_i(\vec{x})$. That is to say, differentiability as defined above implies the existence of the first partial derivatives. Conversely, if f has continuous first partial derivatives near a point $\vec{x}$, then f is differentiable

at $\vec{x}$, as we shall now show. The proof is perhaps forbidding in

notation, but the idea is quite simple: one makes the change from $\vec{x}$

to $\vec{x}+\vec{h}$ in successive steps involving one variable at a time, so that

the definition 1.1(1) and basic property (1), (2) of differentiation

may be invoked. Here is the proof. To establish (5), (6) we put

$\vec{h} = \sum\limits_{i=1}^{k} h_i\vec{e}_i$ and decompose $(\Delta f)(\vec{x},\vec{h})$ as $(\Delta f)(\vec{x},\vec{h}) = f(\vec{x} + \sum h_i\vec{e}_i) - f(\vec{x}) =$

$\sum\limits_{j=1}^{k} \{f(\vec{x} + \sum\limits_{i=j} h_i\vec{e}_i) - f(\vec{x} + \sum\limits_{i=j+1} h_i\vec{e}_i)\}$, where the last term (j=k) is

to be interpreted as $\{f(\vec{x}+h_k\vec{e}_k) - f(\vec{x})\}$. Now $f(\vec{x} + \sum\limits_{i=j} h_i\vec{e}_i) -$

$f(\vec{x} + \sum\limits_{i=j+1} h_i\vec{e}_i) = h_j f_j(\vec{x} + \sum\limits_{i=j+1} h_i\vec{e}_i) + \alpha_j$, where $\alpha_j/h_j \to 0$ as

$h \to 0$, by 1.1(1) and (1), (2). By the assumed continuity of the first

partials, we have $h_j f_j(\vec{x} + \sum\limits_{i=j+1} h_i\vec{e}_i) = h_j f_j(\vec{x}) + h_j\varepsilon_j$, where $\varepsilon_j \to 0$ as

$h_j \to 0$. Therefore $(\Delta f)(\vec{x},\vec{h}) = \sum\limits_{j=1}^{k} h_j f_j(x) + \sum\limits_{j=1}^{k} \{\alpha_j + h_j\varepsilon_j\}$. By means

of the first part of (7) one sees that the error term $\sum\limits_{j=1}^{k} \{\alpha_j + h_j\varepsilon_j\}$

satisfies (6), and the proof is complete.

The following notation is suggestive. Put $d\vec{x} = (dx_1, dx_2, \ldots, dx_k)$

for the increment vector, formerly called $\vec{h}$, and write

$$(df)(\vec{x}, d\vec{x}) = \sum f_i(\vec{x}) dx_i \qquad (8)$$

for the linear term in (5). This quantity, which we emphasize is a

function of $\vec{x}$ and of $d\vec{x}$, is called the differential of the function

$f : \mathbb{R}^k \to \mathbb{R}^1$, and one should note the similarity with the corresponding

formula (3) for a function of one argument. The definition of differentiability may now be given essentially the same formulation, for functions of one or of several arguments: f is differentiable at a point if the increment Δf near the point is approximated by the corresponding differential df within the prescribed accuracy (4) or (6) respectively. For functions of one argument the existence of the derivative is sufficient for this accuracy, and for functions of several arguments we have shown that the existence and continuity of the first partial derivatives is sufficient.

Suppose we have a function $\vec{g} : \mathbb{R}^1 \to \mathbb{R}^k$ with components $g^i : \mathbb{R}^1 \to \mathbb{R}^1$, $i = 1, 2, \ldots, k$. That is, $\vec{g}(t) = (g^1(t), g^2(t), \ldots, g^k(t))$, $t \in \mathbb{R}^1$. If each g^i is differentiable, $(\Delta g^i)(t, h) = h \cdot \dfrac{dg^i}{dt} + \alpha_i(h)$, where $(\alpha_i(h)/h) \to 0$ as $h \to 0$, then $(\Delta \vec{g})(t, h) = h \cdot \dfrac{d\vec{g}}{dt} + \vec{\alpha}(h)$, where $\dfrac{d\vec{g}}{dt} = \left(\dfrac{dg^1}{dt}, \ldots, \dfrac{dg^k}{dt} \right)$ and $\vec{\alpha}(h) = (\alpha_1(h), \ldots, \alpha_k(h))$, and clearly $(\|\vec{\alpha}(h)\|/h) \to 0$ as $h \to 0$. It is therefore a natural extension of the terminology to say in this circumstance (namely, each g^i is differentiable) that $\vec{g}$ is differentiable.

If $\vec{g} : \mathbb{R}^1 \to \mathbb{R}^k$ is differentiable and $f : \mathbb{R}^k \to \mathbb{R}^1$ is differentiable, then the composed function $f \circ \vec{g} : \mathbb{R}^1 \to \mathbb{R}^1$ is differentiable, and

$$\frac{d}{dt}(f \circ \vec{g})(t) = \sum_i f_i(\vec{g}(t)) \cdot \frac{d}{dt} g^i(t). \tag{9}$$

This formula may be guessed from (8), as follows. Putting $\vec{x} = \vec{g}(t)$ there, we have $df(\vec{g}(t), d\vec{g}) = \sum_i f_i(\vec{g}(t)) dg^i$, and dividing now by dt we get (9), more or less. Here is a proof of (9). By the differentiability of $\vec{g}$ we have $\vec{g}(t+h) = \vec{g}(t) + h \cdot \dfrac{d}{dt} \vec{g}(t) + \vec{\alpha}(h)$. Now $(f \circ \vec{g})(t+h) = f(\vec{g}(t+h))$, so by the differentiability of f we have $(f \circ \vec{g})(t+h) = $

$(f \circ \vec{g})(t) + \sum f_i(\vec{g}(t)) \cdot \{h \frac{d}{dt} g^i(t) + \alpha_i(h)\} + \beta(\Delta \vec{g}(t)) =$

$(f \circ \vec{g})(t) + h \sum f_i(\vec{g}(t)) \cdot \frac{d}{dt} g^i(t) + \sum f_i(\vec{g}(t)) \cdot \alpha_i(h) + \beta(\Delta \vec{g}(t))$.   Thus

$\Delta(f \circ \vec{g})(t,h)$ is approximated by $h\{\sum f_i(\vec{g}(t)) \cdot \frac{d}{dt} g^i(t)\}$ with error

$\sum f_i(\vec{g}(t))\alpha_i(h) + \beta(\Delta \vec{g}(t))$.   Now $\frac{1}{h}\sum f_i(\vec{g}(t)) \cdot \alpha_i(h) \to 0$ as $h \to 0$

because $\frac{1}{h}\alpha_i(h) \to 0$ as $h \to 0$;   and $\frac{1}{h}\beta(\Delta \vec{g}(t)) = \frac{\beta(\Delta \vec{g}(t))}{\|\Delta \vec{g}(t)\|} \frac{\|\Delta \vec{g}(t)\|}{h} \to 0$

as $h \to 0$ because the first factor vanishes by the differentiability of

f,   and the second factor equals $\|\frac{d}{dt}\vec{g}(t) + \frac{\vec{\alpha}(h)}{h}\|$,   which approaches

$\|\frac{d}{dt}\vec{g}(t)\|$ as $h \to 0$.   Thus the total error vanishes with the required

rate (4); which is to say $f \circ \vec{g}$ is differentiable and (9) holds.

This formula is an extension to several variables of the chain rule, and we shall refer to it by that name.

1.3   Taylor's Theorem

Given $f: \mathbb{R}^k \to \mathbb{R}^1$,   let $F: \mathbb{R}^1 \to \mathbb{R}^1$ be defined as

$$F(t) = f(\vec{x} + t\vec{h}) \tag{1}$$

for $\vec{x}, \vec{h} \in \mathbb{R}^k$ arbitrary and fixed. Considerations of differentiability and convergence aside, the Taylor series for   F   is

$$F(t) = \sum \frac{t^n}{n!} F^{(n)}(0). \tag{2}$$

By the chain rule 1.2(9) we have

$$F'(t) = \sum_i f_i(\vec{x} + t\vec{h}) \cdot h_i ,$$

$$F''(t) = \sum_{i,j} f_{ij}(\vec{x} + t\vec{h}) h_i h_j ,$$

and in general

$$F^{(n)}(t) = \sum_{i_1, \ldots, i_n} f_{i_1, \ldots, i_n}(\vec{x} + t\vec{h}) h_{i_1} h_{i_2} \cdots h_{i_n} .$$

Therefore

$$F^{(n)}(0) = \sum_{i_1, \ldots, i_n} f_{i_1, \ldots, i_n}(\vec{x}) \cdot h_{i_1} h_{i_2} \cdots h_{i_n} . \tag{3}$$

Putting (3) into (2) and afterwards setting $t = 1$ we find that

$$f(\vec{x} + \vec{h}) = f(\vec{x}) + \sum_i f_i(\vec{x}) h_i + \sum_{ij} f_{ij}(\vec{x}) h_i h_j + \ldots . \tag{4}$$

The chain rule shows, in this case, that if f has continuous partial derivatives up to order n, then F has continuous derivatives up to order n. Therefore, if f has continuous partial derivatives up to order n, then f has an expansion of the form (4) up to order n with an error term, which we write as $R_n(\vec{x}, \vec{h})$. To complete the discussion of (4) we need an estimate of the size of $R_n(\vec{x}, \vec{h})$ when $\vec{h}$ is near $\vec{0}$. It will be sufficient for most of our work to have an estimate for R_1, the error in linear (or first order) approximation; and the method for estimating R_n is clearly indicated by the notationally simpler case of R_1 . The first order Taylor expansion of F is, using the integral form for the remainder, [1]

$$F(t) = F(0) + tF'(t) + \int_0^t (t-s) F''(s) ds. \tag{5}$$

Setting $t=1$ we have for f the expansion

$$f(\vec{x}+\vec{h}) = f(\vec{x}) + \sum_i f_i(\vec{x}) h_i + \int_0^1 (1-s) F''(s) ds . \tag{6}$$

Thus $R_1(\vec{x},\vec{h}) = \int_0^1 (1-s) F''(s) ds$. Since $F''(s) = \sum_{i,j} f_{ij}(\vec{x}+s\vec{h}) h_i h_j$ we have

$$|R_1(\vec{x},\vec{h})| \leq \{\mathrm{Max}|f_{ij}|\} \cdot \left(\sum_{i,j} |h_i h_j|\right) \cdot \frac{1}{2}, \tag{7}$$

the factor $\frac{1}{2}$ coming from the integration of $(1-s)$. Using the inequality 1.2(7) applied to $\vec{h}$ we have $\sum_{i,j} |h_i h_j| \leq k^2 \|\vec{h}\|^2$; whence, writing $M = \mathrm{Max}|f_{ij}|$, we get

$$|R_1(\vec{x},\vec{h})| \leq \frac{1}{2} M \cdot k^2 \cdot \|\vec{h}\|^2, \tag{8}$$

which implies finally that

$$\frac{R_1(\vec{x},\vec{h})}{\|\vec{h}\|} \to 0 \quad \text{as} \quad \vec{h} \to \vec{0} . \tag{9}$$

Thus the error term after linear approximation vanishes faster than the

---

[1] See Courant [1], volume I, p. 323.

length $\|\vec{h}\|$ of the increment vector $\vec{h}$ as $\vec{h} \to \vec{0}$, in the sense (9), for a function f of several variables having continuous partial derivatives up to order 2. In general, when f has continuous partials up to order n+1, the error $R_n(\vec{x},\vec{h})$ vanishes faster than $\|\vec{h}\|^n$, in the sense that

$$\frac{R_n(\vec{x},\vec{h})}{\|\vec{h}\|^n} \to 0 \quad \text{as} \quad \vec{h} \to \vec{0} . \tag{10}$$

This is established by the same method as was (9). The existence of the expansion (4) up to order n with error term R_n satisfying (10) constitutes Taylor's theorem for a function $f : \mathbb{R}^k \to \mathbb{R}^1$ of k variables.

Consider now a k-vector-valued function $f : \mathbb{R}^k \to \mathbb{R}^k$ of a k-vector argument. Let $f^\nu : \mathbb{R}^k \to \mathbb{R}^1$, $\nu = 1, 2, \ldots, k$ be the components of $\vec{f}$. That is,

$$\vec{f}(\vec{x}) = (f^1(\vec{x}), f^2(\vec{x}), \ldots, f^k(\vec{x})) .$$

A Taylor expansion for $\vec{f}$ may be assembled out of the Taylor expansions for the components f^ν. The terms become complicated as one goes to higher orders, but we shall have need only of the first order term, which is both simple and pretty. We write out (4) for each f^ν, and find that the first order approximant to $\vec{f}(\vec{x}+\vec{h})$ is the vector whose νth component is $\sum_i f_i^\nu(\vec{x}) h_i$. This suggests forming the matrix whose entries are the first order partials of the component functions f^ν. This matrix is denoted $d\vec{f}(\vec{x})$ and known as the differential of $\vec{f}$.

By definition, then,

$$d\vec{f}(\vec{x}) \;=\; ((\; f^{\nu}_{i}(\vec{x}) \;)), \qquad\qquad (11)$$

where the upper (component) index is the row index, and the lower
(derivative) index is the column index. We then have

$$\vec{f}(\vec{x}+\vec{h}) \;=\; \vec{f}(\vec{x}) + d\vec{f}(\vec{x})\vec{h} + \ldots \qquad\qquad (12)$$

where the second term is the result of applying the matrix $d\vec{f}(\vec{x})$ to
the vector $\vec{h}$. Thus, up to first order we have, for functions
$\vec{f}: \mathbb{R}^{k} \to \mathbb{R}^{k}$, the ordinary one-variable linear approximation formula
decorated with arrows.

The following terminology is standard. A function $f: \mathbb{R}^{k} \to \mathbb{R}^{1}$
is called a <u>scalar</u> <u>field</u>, and may be thought of as attaching a scalar
$f(\vec{x})$ to each point $\vec{x}$ of $\mathbb{R}^{k}$. A function $\vec{f}: \mathbb{R}^{k} \to \mathbb{R}^{k}$ is called a
<u>vector</u> <u>field</u>, and may be thought of as attaching a vector $\vec{f}(\vec{x})$ to
each point $\vec{x}$ of $\mathbb{R}^{k}$. In this terminology (4) is the Taylor expansion
of a scalar field f, and we have just seen the expansion to first
order (12) of a vector field $\vec{f}$.

Concerning the vector field differentials (11), we have the
important fact that the differential of a composition of two vector
fields is the matrix product of their differentials. In detail this
is the assertion that if the vector fields $\vec{f}$ and $\vec{g}$ are differentiable
then so is $\vec{f} \circ \vec{g}$ and

$$d(\vec{f} \circ \vec{g})(\vec{x}) \;=\; d\vec{f}(\vec{g}(x)) \cdot d\vec{g}(x) \;.$$ (13)

For the proof, we observe first that $(\vec{f} \circ \vec{g})^{\vee}(\vec{x}) = f^{\vee}(\vec{g}(\vec{x}))$, so that $\dfrac{\partial}{\partial x_{\mu}} (\vec{f} \circ \vec{g})^{\vee}(\vec{x}) = \sum_{\ell} f^{\vee}_{\ell}(\vec{g}(\vec{x})) g^{\ell}_{\mu}(\vec{x})$. By the definition (11) of differentials for vector fields, together with the definition (row-by-column rule) of product for matrices, the last equation is precisely (13).

Chapter 2
Differential forms

2.1 Line Integrals

A curve in $\mathbb{R}^k$ is specified by giving a function $\vec{\gamma} : \mathbb{R}^1 \to \mathbb{R}^k$.
Then $\vec{\gamma}(t)$ is that point on the curve corresponding to the parameter
value t.

Let $\vec{\gamma} : \mathbb{R}^1 \to \mathbb{R}^k$ be defined for $a \leq t \leq b$. Subdivide $[a,b]$
as $a = t_0 < t_1 < \ldots < t_n = b$,
and let $\ell_1, \ell_2, \ldots, \ell_n$ be the
corresponding chords of the curve
(see figure). Now

$$\|\ell_i\|^2 = \sum_j \left(\gamma^j(t_i) - \gamma^j(t_{i-1}) \right)^2$$

$$= \sum_j \left\{ \frac{\gamma^j(t_i) - \gamma^j(t_{i-1})}{(t_i - t_{i-1})} \right\}^2 \left(t_i - t_{i-1} \right)^2$$

$$= \sum_j \{ \dot{\gamma}^j(t_{i-1}) + \eta_j \}^2 (t_i - t_{i-1})^2 ,$$

where $\eta_j \to 0$ as $\Delta t_i = t_i - t_{i-1} \to 0$, and $\dot{\gamma}^j = \frac{d}{dt} \gamma^j$. The length

L of the curve is defined as $L = \lim_{\Delta t_i \to 0} \sum_i \| \ell_i \|$, whence

$$L = \lim_{\Delta t_i \to 0} \sum_i \left(\sqrt{\sum_j (\dot{\gamma}^j + \eta_j)^2} \right) \cdot \Delta t_i$$

$$= \int_a^b \sqrt{\sum_j (\dot{\gamma}^j)^2} \; dt \; ,$$

and the length $s(t)$ of the curve from $\vec{\gamma}(a)$ to $\vec{\gamma}(t)$ is

$$s(t) = \int_a^t \sqrt{\sum_j (\dot{\gamma}^j(u))^2} \; du \; . \tag{1}$$

Therefore

$$\frac{ds}{dt} = \sqrt{\sum_j (\dot{\gamma}^j(t))^2} \; . \tag{2}$$

The vector $(\dot{\gamma}^1(t), \dot{\gamma}^2(t), \ldots, \dot{\gamma}^k(t))$ is tangent to the curve at the

point $\vec{\gamma}(t)$. We define $\frac{d}{dt} \vec{\gamma}(t)$ as this vector. Then (2) becomes

$\frac{ds}{dt} = \| \frac{d}{dt} \vec{\gamma}(t) \|$, or

$$ds = \| \frac{d}{dt} \vec{\gamma}(t) \| \; dt \; . \tag{3}$$

This expression is called the <u>element of arc length</u>, or the <u>line element</u>.

Suppose we change to a new variable u, and let t = φ(u), u = φ^{-1}(t) denote the functional relation between the two variables. We have

$$ds = \sqrt{\sum_j (\dot{\gamma}^j(t))^2 \phi'(u)^2} \; \frac{dt}{\phi'(u)} \; ; \qquad (4)$$

but $\dot{\gamma}^j(t) \phi'(u) = \frac{d}{du} \gamma^j(\phi(u))$ by the chain rule, and dt = φ'(u)du, du = $\frac{dt}{\phi'(u)}$ = $\frac{d}{dt} \phi^{-1}(t) dt$; so that (4) is

$$ds = \left\| \frac{d}{du} (\vec{\gamma} \circ \phi)(u) \right\| du . \qquad (5)$$

That is to say, the form of the line element is preserved under a change of variable.

We may apply this procedure to the arc length along the curve. That is, we may change from the given variable t to the new variable s defined by (1). Then $\phi^{-1}(t) = \int_a^t \left\| \frac{d}{du} \vec{\gamma}(u) \right\| du = s$, from which one can in principle determine φ, and then $\left\| \frac{d}{ds}(\vec{\gamma} \circ \phi)(s) \right\| ds = ds$. This shows that the length of the tangent vector (the speed of traversing the curve, so to speak) is at each point equal to 1 if the variable in which the curve is expressed is its own arc length. And the converse of this statement comes at once from (3). That is, if the tangent vector is everywhere of length 1, then the given variable differs by a constant from s .

By the line integral $\int_C f$ of a scalar field $f : \mathbb{R}^k \to \mathbb{R}^1$ along a curve C we mean the limit of sums of values of f at points of

C times the line elements at those points. This is nothing but the curvilinear version of the ordinary integral of a function $f : \mathbb{R}^1 \to \mathbb{R}^1$ along the X-axis. If the curve is given as $\vec{\gamma} : \mathbb{R}^1 \to \mathbb{R}^k$ through the variable t, then

$$\int_C f = \int_{t_0}^{t_1} (f \circ \vec{\gamma})(t) \| \frac{d}{dt} \vec{\gamma}(t) \| dt , \qquad (6)$$

where $\vec{\gamma}(t_0), \vec{\gamma}(t_1)$ are the endpoints of C .

By the line integral $\int_C \vec{f}$ of a vector field $\vec{f} : \mathbb{R}^k \to \mathbb{R}^k$ along a curve C we mean something rather different; namely, we mean the limit of sums of the component of $\vec{f}$ tangent to C at each point of C times the line element at the point. The familiar work integral of physics is an example, and brings out the idea: if $\vec{f}$ is a field of force, the element of work done on a test particle by the field in moving it along the curve is the component of $\vec{f}$ along the curve times the distance (line element) through which it acts. If C is given as $\vec{\gamma} : \mathbb{R}^1 \to \mathbb{R}^k$ with variable t then we have

$$\int_C \vec{f} = \int_{t_0}^{t_1} (\vec{f} \circ \vec{\gamma})(t) \cdot (\frac{d}{dt} \vec{\gamma}(t)) dt . \qquad (7)$$

The line element $\| \frac{d}{dt} \vec{\gamma}(t) \| dt$ seems to be absent, but it is really there: writing $\dot{\vec{\gamma}}$ for $\frac{d}{dt} \vec{\gamma}$, the component of f along C is $(\vec{f} \circ \gamma) \cdot (\dot{\vec{\gamma}} / \|\dot{\vec{\gamma}}\|)$, so that $(\vec{f} \circ \gamma) \cdot (\dot{\vec{\gamma}} / \|\dot{\vec{\gamma}}\|) \|\dot{\vec{\gamma}}\| dt$, which should appear in (7), equals what does appear there.

Expressing the vectors in (7) in components and evaluating the dot product thereby, we find

$$\int_C \vec{f} = \int_{t_0}^{t_1} \sum_i (f^i \circ \vec{\gamma})(t) \frac{d}{dt} \gamma^i(t) \, dt \qquad (8)$$

(because $(\vec{f} \circ \vec{\gamma})^i = f^i \circ \vec{\gamma})$. This suggests the following abstract

notation for $\int_C \vec{f}$:

$$\int_C \vec{f} = \int_C \{\sum_i f^i dx_i\} . \qquad (9)$$

This employes the notation of 1.2(8). The right side of (9) could also

be written $\int_C \vec{f} \cdot d\vec{x}$. The advantage of the abstract notation (9) is

that one is not commited to any particular choice of presentation of C.

Of course if a parametric presentation $\vec{\gamma} : \mathbb{R}^1 \to \mathbb{R}^k$ is given for C

then (9) transforms itself at once back into (7).

Note that $\int_C \vec{f}$ is an _oriented_ object: if the orientation of

the curve (direction of increasing variable) is reversed, then $\int_C \vec{f}$

will change its sign, because the tangent vector will then change its

sign everywhere on C .

2.2 One-Forms

In elementary calculus the object dx is called the differential

of the independent variable x and has two meanings: it is the generic

increment of x and yields, through the formula $df = f'(x) dx$, the

first order approximation to the corresponding increment of a function

f ; secondly, it indicates the dummy variable in integration. The

vectorial generalization of this object would be the generic increment vector $\vec{dx} = (dx_1, dx_2, \ldots, dx_k)$ introduced in 1.2(8). It determines, through the formula $df = \sum_i f_i dx_i$, the first order approximation to the corresponding increment of a scalar field f (see 1.2(8), 1.3(4)); and it indicates the integration variables in the abstract form $\int_C \vec{g} = \int_C \{\sum g^i dx_i\}$ of the line integral of a vector field $\vec{g}$. Note that the form $\sum_i f_i dx_i$ resembles $\sum_i g^i dx_i$ and is a specialization of it: in the former the components of the vector field are in fact the partial derivatives of a given scalar field. We call an expression of the general form $\sum_i g^i dx_i$, where $\vec{g} = (g^1, \ldots, g^k)$ is an arbitrary vector field and $\vec{dx} = (dx_1, \ldots, dx_k)$ is the generic increment vector, a differential form of degree one, or briefly a one-form, in k variables.

Our definition of one-forms is expressed in terms of the standard Cartesian coordinate system in $\mathbb{R}^k$. This is an arbitrary choice, as we shall indicate later on (Section 2.4). We are going to introduce an algebraic structure into the set of forms, which yields in particular a natural concept of forms of higher degree, all this in Cartesian coordinates; then we shall inquire how this apparatus behaves under a passage to new coordinates of arbitrary type; and finally we will observe that all essential desiderata of the apparatus persist under coordinate changes.

As the first step in this program we now show how the set of one-forms in Cartesian coordinates in $\mathbb{R}^k$ may be seen as a linear space (that is, a set of objects which can be added like vectors and which admit a distributive "scalar" multiplication). To this end, think of $dx_1, dx_2, \ldots, dx_k$ as abstract objects which are to be the basis of a linear space whose scalars are the set of all scalar fields. Then the

generic element of this space is a linear combination $\sum_i g^i dx_i$ of

the basis elements dx_i with scalar field coefficients $g^i : \mathbb{R}^k \to \mathbb{R}^1$.

Let

$$\sigma = \sum_i f^i dx_i , \qquad \tau = \sum_i g^i dx_i$$

be two such "vectors". Define the sum $\sigma + \tau$ by

$$\sigma + \tau = \sum_i (f^i + g^i) dx_i . \tag{1}$$

This addition is like that of ordinary vectors: it is commutative,

$\sigma + \tau = \tau + \sigma$;   it is associative, $(\sigma + \tau) + \theta = \sigma + (\tau + \theta)$ where

$\theta = \sum_i h^i dx_i$,   say; and there is the zero "vector" $0 = \sum_i 0\, dx_i$.   If

h is a scalar field, we define the scalar multiplication $h\sigma$ by

$$h\sigma = \sum_i (hf^i) dx_i . \tag{2}$$

That is, the ith component of $h\sigma$ is h times the ith component

of σ . Note that hf^i is the numerical product of the functions and

not a functional composition. That is, given $f^i : \mathbb{R}^k \to \mathbb{R}^1$ and

$h : \mathbb{R}^k \to \mathbb{R}^1$, $hf^i(\vec{x}) = h(\vec{x}) f^i(\vec{x})$,   whence $hf^i : \mathbb{R}^k \to \mathbb{R}^1$ is again a

scalar field, as required in (2). One sees immediately that the scalar

multiplication (2) distributes across the "vector" addition (1),

$$h(\sigma + \tau) = h\sigma + h\tau , \tag{3}$$

this being precisely the fact that scalar-field product distributes across scalar-field addition. It is likewise clear, since scalar-field product distributes across scalar-field addition, that $(f+g)\sigma =$ $f\sigma+g\sigma$, as is the case for ordinary scalars (real numbers) and ordinary vectors.

We denote by $\mathbb{S}$ the set of all scalar fields, and by Λ^1 the space of all one-forms in Cartesian coordinates in $\mathbb{R}^k$. We have now established that Λ^1 is a linear space with $\mathbb{S}$ as scalars, or, briefly, that Λ^1 is a linear space over $\mathbb{S}$, under the operations (1) and (2).

Let $\mathbb{R}$ denote the set of real numbers. We may regard $\mathbb{R}$ as a subset of $\mathbb{S}$ in the following natural sense: each $t \in \mathbb{R}$ determines the <u>constant scalar field</u> $f_t \in \mathbb{S}$, where

$$f_t(\vec{x}) \equiv t .$$

With this convention the product of $t \in \mathbb{R}$ and $\sigma \in \Lambda^1$ is defined by (2): $t\sigma = f_t\sigma = \sum_i (f_t g^i) dx_i = \sum_i (tg^i) dx_i$. For example

$$-\sigma = \sum_i (-g^i) dx_i . \tag{4}$$

2.3 Wedge Product

We now construct a second linear space over $\mathbb{S}$. Its basis is the set of all pairs of the basic one-forms $dx_1,\ldots,dx_k$. We write the generic pair as $dx_i \wedge dx_j$, read dx_i <u>wedge</u> dx_j . The generic element of the new space has the form $\sum_{ij} g^{ij} dx_i \wedge dx_j$, where $g^{ij} : \mathbb{R}^k \to \mathbb{R}^1$

is a doubly indexed set of scalar fields, k^2 in number. Addition

and $\mathbb{S}$-multiplication are defined in this space just as in Λ^1: if

$\sigma = \sum_{ij} f^{ij} dx_i \wedge dx_j$ and $\tau = \sum_{ij} g^{ij} dx_i \wedge dx_j$, then $\sigma + \tau =$

$\sum_{ij} (f^{ij} + g^{ij}) dx_i \wedge dx_j$ and if $h \in \mathbb{S}$, then $h\sigma = \sum_{ij} (hf^{ij}) dx_i \wedge dx_j$.

Now it happens that this space is too big for our purposes, and we

contract it by imposing the following <u>axiom</u>: for all i, j,

$$dx_i \wedge dx_j = -dx_j \wedge dx_i . \tag{1}$$

It follows that for all i

$$dx_i \wedge dx_i = 0 , \tag{2}$$

for $dx_i \wedge dx_i = -dx_i \wedge dx_i$, whence $2 dx_i \wedge dx_i = 0$, and this entails

(2). The contracted space is denoted Λ^2, and the elements of Λ^2

are called the <u>two-forms</u> in Cartesian coordinates in $\mathbb{R}^k$. The basis

of Λ^2 consists of all linearly independent pairs $dx_i \wedge dx_j$. By (1),

$dx_i \wedge dx_j$ and $dx_j \wedge dx_i$ are dependent, and by (2) the "squares"

$dx_i \wedge dx_i$ are excluded. Thus the number of basic two-forms is the number

of pairs which may be chosen from the set $\{dx_1, \ldots, dx_k\}$ of k basic

one-forms. This number is $\binom{k}{2}$, the number of combinations of k things

taken 2 at a time.[1] A convenient and customary way of listing this

basis is

$$dx_i \wedge dx_j, \quad i < j, \quad i, j = 1, 2, \ldots, k . \tag{3}$$

---

[1] See Page (v) for the formula for $\binom{k}{j}$, the number of combinations of k things j at a time, $0 \le j \le k$.

The restriction $i < j$ prevents the duplication and redundance which are to be excluded by (1) and (2). In $\mathbb{R}^3$, for example, one has the $\binom{3}{2} = 3$ basis elements $dx_1 \wedge dx_2$, $dx_1 \wedge dx_3$, and $dx_2 \wedge dx_3$. The generic two-form in $\mathbb{R}^k$ is

$$\sum_{i<j} g^{ij} dx_i \wedge dx_j \ , \tag{4}$$

where $g^{ij} : \mathbb{R}^k \to \mathbb{R}^1$, $i < j$, are scalar fields, $\binom{k}{2}$ in number, and the summation is over all pairs $i,j = 1,2,\ldots,k$ such that $i < j$. We shall presently see that two-forms are to surface integrals what one-forms are to line integrals.

We wish now to interpret wedge as a binary operation, to be called __wedge product__, on one-forms. To see how this is to be done, we consider a speculative "calculation" in $\mathbb{R}^2$. Let $\sigma = f^1 dx_1 + f^2 dx_2$, $\tau = g^1 dx_1 + g^2 dx_2$ be one-forms in $\mathbb{R}^2$. What is the "product"

$$\sigma \wedge \tau = (f^1 dx_1 + f^2 dx_2) \wedge (g^1 dx_1 + g^2 dx_2)$$

to mean? If wedge as a binary operation were distributive with respect to the addition in Λ^1 and homogeneous with respect to its $\mathbb{S}$-multiplication, then $\sigma \wedge \tau$ would be the two-form $(f^1 g^2 - f^2 g^1) dx_1 \wedge dx_2$, there being only one term instead of four by virtue of (1) and (2). We therefore add to (1) two additional __axioms__ for wedge, as follows:

$$dx_i \wedge (f dx_j) = (f dx_i) \wedge dx_j = f(dx_i \wedge dx_j) \ , \tag{5}$$

$$dx_i \wedge (\textstyle\sum_j g^j dx_j) = \textstyle\sum_j dx_i \wedge (g^j dx_j) \ ; \tag{6}$$

the first is homogeneity with respect to $-multiplication, and the second is distributivity with respect to Λ^1-addition. Note that (5) applied to (6) yields

$$dx_i \wedge (\sum_j g^j dx_j) = \sum_j g^j (dx_i \wedge dx_j) , \qquad (7)$$

the "product" is a two-form. We now define the wedge product of one-forms to be the binary relation given by wedge, subject to the three axioms (1), (5), (6). Clearly this product assigns a two-form to each pair of one-forms, in symbols $\Lambda^1 \wedge \Lambda^1 \subset \Lambda^2$; it is associative, distributive with respect to Λ^1-addition, and homogeneous with respect to $-multiplication; but it is <u>not commutative</u> by virtue of (1). In (5) and (6) we have put the factor dx_i on the left side. The corresponding statement with the factor on the right side is given at once by (1).

Here are a few sample wedge products in $\mathbb{R}^3$.

(i) $\qquad (f^1 dx_1 + f^2 dx_2 + f^3 dx_3) \wedge dx_3 = f^1 dx_1 \wedge dx_3 + f^2 dx_2 \wedge dx_3 .$

Note the vanishing of the third term, by (2).

(ii) $\qquad dx_2 \wedge (f^1 dx_1 + f^2 dx_2 + f^3 dx_3) = -f^1 dx_1 \wedge dx_2 + f^3 dx_2 \wedge dx_3 .$

Note the sign change, as required by (1).

Of course we are obliged to show that Axioms (1), (5), (6) are consistent. We defer this to a later point where it will be more conveniently done.[1] We shall see that the wedge product is a remarkable generalization of the familiar vector product of vectors in $\mathbb{R}^3$.

_____

(1) See the Appendix, page 144.

The formula for the wedge product of one-forms in $\mathbb{R}^k$, which is to say, the general form of our speculative calculation above, is:

$$\left(\sum_i f^i dx_i\right) \wedge \left(\sum_j g^j dx_j\right) = \sum_{i<j} \left(f^i g^j - f^j g^i\right) dx_i \wedge dx_j , \qquad (8)$$

where the notation $\sum_{i<j}$ is that introduced in (4): summation over all pairs $i,j = 1,2,\ldots,k$ subject to the restriction $i < j$. The minus sign on the right side of (8) arises from (1) when the wedge products of the basic one-forms dx_i are put in the order (3) of the basic two-forms. There is nothing difficult about (8), but it should be carefully pondered and well absorbed, for it contains the basic arithmetic of differential forms. Here is (8) for $\mathbb{R}^3$.

$$(f^1 dx_1 + f^2 dx_2 + f^3 dx_3) \wedge (g^1 dx_1 + g^2 dx_2 + g^3 dx_3)$$

$$= (f^1 g^2 - f^2 g^1) dx_1 \wedge dx_2 + (f^1 g^3 - f^3 g^1) dx_1 \wedge dx_3 + (f^2 g^3 - f^3 g^2) dx_2 \wedge dx_3 . \qquad (9)$$

This should remind the reader of the vector product $\vec{f} \times \vec{g}$ of $\vec{f} = (f^1, f^2, f^3)$ and $\vec{g} = (g^1, g^2, g^3)$ in $\mathbb{R}^3$.

The foregoing construction of $\wedge^2$ generalizes as follows. Given $\mathbb{R}^k$, the set $\mathbb{S}$ of scalar fields in $\mathbb{R}^k$, and the generic increment vector $\vec{dx} = (dx_1, \ldots, dx_k)$, for each integer r we form the linear space over $\mathbb{S}$ whose basis is the set of all r-tuples of the basic one-forms $dx_1, \ldots, dx_k$. We write the generic r-tuple as

$dx_{i_1} \wedge dx_{i_2} \wedge \ldots \wedge dx_{i_r}$, where $i_1, i_2, \ldots, i_r$ is any set of r integers chosen from the set $\{1, 2, \ldots, k\}$. Addition and $\mathbb{S}$-multiplication in this space are componentwise, in exact analogy with 2.2(1), 2.2(2). We contract this space by imposing upon r-tuples the analog of (1): the interchange of adjacent terms in $dx_{i_1} \wedge \ldots \wedge dx_{i_r}$ changes its sign; and generally, m such interchanges cause the r-tuple to be multiplied by $(-1)^m$. For example,

$$dx_{i_2} \wedge dx_{i_1} \wedge dx_{i_3} \wedge \ldots \wedge dx_{i_r} = -dx_{i_1} \wedge dx_{i_2} \wedge \ldots \wedge dx_{i_r},$$

$$dx_{i_3} \wedge dx_{i_1} \wedge dx_{i_2} \wedge \ldots \wedge dx_{i_r} = +dx_{i_1} \wedge dx_{i_2} \wedge \ldots \wedge dx_{i_r}.$$

It follows, just as does (2) from (1), that an r-tuple containing a duplication vanishes. This is because by successive interchanges the duplicate terms can be brought adjacent; and therefore, by the general form of (1) just stated above, the product will equal its negative and so vanish. In particular, an r-tuple $dx_{i_1} \wedge \ldots \wedge dx_{i_r}$ with $r > k$ terms must vanish because it must contain a duplication. Therefore the independent r-tuples correspond to subsets of size $r \leq k$ chosen from $\{1,2,\ldots,k\}$ and may be listed as

$$dx_{i_1} \wedge dx_{i_2} \wedge \ldots \wedge dx_{i_r}, \quad 1 \leq i_1 < i_2 < \ldots < i_r \leq k, \tag{10}$$

as in (3). The contracted $\mathcal{S}$-space with basis (10) is denoted $\wedge^r$, and its elements are the __r-forms__ in Cartesian coordinates in $\mathbb{R}^k$. The generic r-form is

$$\sum_{i_1 < \ldots < i_r} f^{i_1,\ldots,i_r} \, dx_{i_1} \wedge \ldots \wedge dx_{i_r}, \tag{11}$$

where $f^{i_1,\ldots,i_r} : \mathbb{R}^k \to \mathbb{R}^1$ are scalar fields and the summation extends over all $i_1,\ldots,i_r$ such that $1 \le i_1 < i_2 < \ldots < i_r \le k$. Clearly the

number of independent basic r-forms (10) is $\binom{k}{r}$, and we shall say

therefore that Λ^r has dimension $\binom{k}{r}$. The hierarchy of form spaces

Λ^r stops at r=k, or as one may say, $\Lambda^r = 0$ for $r > k$, for, as we

have noted above, an r-tuple $dx_{i_1} \wedge \ldots \wedge dx_{i_r}$ with $r > k$ terms vanishes.

The dimension of Λ^k is $\binom{k}{k} = 1$, and indeed, there is only one k-tuple

$dx_1 \wedge \ldots \wedge dx_k$ (apart from interchanges, which only change its sign), so

the generic k-form in $\mathbb{R}^k$ is $f\, dx_1 \wedge \ldots \wedge dx_k$, where $f : \mathbb{R}^k \to \mathbb{R}^1$ is

any scalar field. Our construction of the form spaces leaves open the

bottom of the hierarchy, r=0. We complete the scheme by regarding

scalar fields as <u>zero-forms</u>, and setting $\Lambda^0 = \mathfrak{S}$. Note that $\binom{k}{0} = 1$.

For the basis of Λ^0 we make the constant scalar field $f_1(x) \equiv 1$.

The interpretation of wedge as a binary operation (product) on

pairs of forms also generalizes. Let r,s be integers with $1 \le r \le k$,

$1 \le s \le k$. We postulate the analogs of (5) and (6) for the wedge combina-

tion of the basic s-forms with elements of Λ^r: namely, that it is

homogeneous with respect to $\mathfrak{S}$-multiplication, and that it is distributive

with respect to the addition in Λ^r. Matters being so, the wedge product

of an r-form and an s-form is an (r+s)-form; or in symbols $\Lambda^r \wedge \Lambda^s \subset \Lambda^{r+s}$.

As an example, let $\sigma = \sum_i f^i dx_i$, $\tau = \sum_j g^j dx_j$ be one-forms in $\mathbb{R}^k$.

Then we have

$$\sigma \wedge \tau = \sum_{i<j} (f^i g^j - f^j g^i)\, dx_i \wedge dx_j,$$

which is (8). Observe that $\sigma \wedge \tau = -\tau \wedge \sigma$. Next let $\sigma = \sum_i f^i dx_i \in \Lambda^1$,

$\tau = \sum_{i<j} g^{ij} dx_i \wedge dx_j \in \Lambda^2$. be given. Then

$$\sigma \wedge \tau = \sum_{i<j<k} (f^i g^{jk} - f^j g^{ik} + f^k g^{ij}) \, dx_i \wedge dx_j \wedge dx_k \, , \tag{12}$$

a three-form. The signs come, as in (8), from the interchange rule (1)

in its general form. For instance, $f^2 g^{13} dx_2 \wedge dx_1 \wedge dx_3 = -f^2 g^{13} dx_1 \wedge dx_2 \wedge dx_3$

because one interchange is required to put the indices in $dx_2 \wedge dx_1 \wedge dx_3$

into increasing order. Observe in this case that $\sigma \wedge \tau = \tau \wedge \sigma$.

The anti-commutation $\sigma \wedge \tau = -\tau \wedge \sigma$ of (8) and the commutation

$\sigma \wedge \tau = \tau \wedge \sigma$ of (12) are both instances of the following general commu-

tation rule for the wedge product.

<u>Theorem</u>. If $\sigma \in \Lambda^r$ and $\tau \in \Lambda^s$, then

$$\sigma \wedge \tau = (-1)^{rs} \tau \wedge \sigma . \tag{13}$$

Proof: Let $\sigma = \sum_{i_1 < \ldots < i_r} f^{i_1, \ldots, i_r} dx_{i_1} \wedge \ldots \wedge dx_{i_r} \, ,$

$\tau = \sum_{j_1 < \ldots < j_r} g^{j_1, \ldots, j_s} dx_{j_1} \wedge \ldots \wedge dx_{j_s} \, .$ Since wedge product distributes

across summation, it suffices to verify (13) for a generic single-term

product

$$(f^{i_1, \ldots, i_r} dx_{i_1} \wedge \ldots \wedge dx_{i_r}) \wedge (g^{j_1, \ldots, j_s} dx_{j_1} \wedge \ldots \wedge dx_{j_s}) \tag{14}$$

in $\sigma \wedge \tau$. Let us denote these single-terms (monomials) by α, β

respectively. The product (14) is then $\alpha \wedge \beta$. If the index sets

$\{i_1, \ldots, i_r\}$ and $\{j_1, \ldots, j_s\}$ overlap, then $\alpha \wedge \beta = \beta \wedge \alpha = 0$ because

there will then be a duplication in the product

$dx_{i_1} \wedge \ldots \wedge dx_{i_r} \wedge dx_{j_1} \wedge \ldots \wedge dx_{j_s}$. Therefore such terms contribute 0 to both sides of (13). In the contrary case (that the index sets are disjoint), $\alpha \wedge \beta$ and $\beta \wedge \alpha$ are monomial (r+s)-forms,

$$\alpha \wedge \beta = f^{i_1, \ldots, i_r} g^{j_1, \ldots, j_s} dx_{i_1} \wedge \ldots \wedge dx_{i_r} \wedge dx_{j_1} \wedge \ldots \wedge dx_{j_s},$$

$$\beta \wedge \alpha = f^{i_1, \ldots, i_r} g^{j_1, \ldots, j_s} dx_{j_1} \wedge \ldots \wedge dx_{j_s} \wedge dx_{i_1} \wedge \ldots \wedge dx_{i_r}.$$

To bring the differentials in $\beta \wedge \alpha$ into the order they have in $\alpha \wedge \beta$ requires bringing r dx_i's past s dx_j's, or in all $r \cdot s$ interchanges, whence, by (1) in its general form, $\beta \wedge \alpha = (-1)^{rs} \alpha \wedge \beta$. We have shown that (13) holds for every non-zero term in $\sigma \wedge \tau$, and hence we have established (13), QED.

Corollary. If $\sigma \in \Lambda^r$, r odd, then $\sigma \wedge \sigma = 0$.

Proof: $\sigma \wedge \sigma = (-1)^{r^2} \sigma \wedge \sigma$, and $(-1)^{r^2} = -1$ if r is odd, whence $\sigma \wedge \sigma = -\sigma \wedge \sigma = 0$, QED.

Exercise. Show by examples that for forms σ of even degree, $\sigma \wedge \sigma$ may vanish but need not.

2.4 Change of Coordinates

A vector field $\vec{f} : \mathbb{R}^k \to \mathbb{R}^k$ may be interpreted either as a coordinate change in $\mathbb{R}^k$ or as a transformation (motion) of the points of $\mathbb{R}^k$. Denote as usual the components of $\vec{f}$ in the fixed Cartesian coordinate system by f^i:

$$\vec{f}(\vec{x}) = \sum_i f^i(\vec{x}) \vec{e}_i, \tag{1}$$

where $\vec{e}_1, \vec{e}_2, \ldots, \vec{e}_k$ are the mutually orthogonal unit vectors of the Cartesian system (see page 2). If $\vec{f}$ is regarded as a transformation, so that $\vec{f}(\vec{x})$ is interpreted as the point to which $\vec{f}$ moves $\vec{x}$, then (1) gives the Cartesian coordinates $(f^1(\vec{x}), \ldots, f^k(\vec{x}))$ of the image point. If $\vec{f}$ is regarded as a coordinate change, one thinks of the points of $\mathbb{R}^k$ as intrinsic and immobile but capable of many descriptions; and then (1) introduces a new description according to which $\vec{x}$ is labeled $(f^1(\vec{x}), \ldots, f^k(\vec{x}))$. It is customary to abbreviate the new labels, say $y_i = f^i(\vec{x})$, and write $\vec{y} = (y_1, \ldots, y_k)$. This is an abuse of notation because $(y_1, \ldots, y_k)$ is not, in the coordinate change interpretation, the set of Cartesian coordinates of a point $\vec{y}$ in $\mathbb{R}^k$, whereas the notation gives that appearance.

In the coordinate change interpretation the new coordinate "lines", the loci of points generated by setting all but one y_i equal to a constant, are typically curves, and the new coordinates are generally known as curvilinear coordinates.

As an example we consider polar coordinates in $\mathbb{R}^2$. Let (x_1, x_2) be the Cartesian system coordinates, let (y_1, y_2) be the polar system coordinates, where $0 \le y_1 < \infty$, $0 \le y_2 < 2\pi$, and let $\vec{f} : \mathbb{R}^2 \to \mathbb{R}^2$ effect the change from Cartesian to polar coordinates. Then

$$y_1 = f^1(\vec{x}) = \sqrt{x_1^2 + x_2^2}$$

$$(2)$$

$$y_2 = f^2(\vec{x}) = \text{Arctan} \, \frac{x_2}{x_1}.$$

The differentials (one-forms) of f^1, f^2 in Cartesian coordinates are

$$df^1 = \frac{x_1}{\sqrt{x_1^2 + x_2^2}} dx_1 + \frac{x_2}{\sqrt{x_1^2 + x_2^2}} dx_2 ,$$

(3)

$$df^2 = \frac{-x_2}{x_1^2 + x_2^2} dx_1 + \frac{x_1}{x_1^2 + x_2^2} dx_2 ,$$

as one sees by partial differentiation $(df^1 = f_1^1 dx_1 + f_2^1 dx_2 ,$ $df^2 = f_1^2 dx_1 + f_2^2 dx_2)$. If we set y_2 constant, say $y_2 = c$, the locus of points of $\mathbb{R}^2$ one gets is a straight line whose Cartesian equation is (by (2))

$$x_2 = (\tan c) x_1 .$$

(4)

Its slope thus varies with c, and one gets in this way the familiar polar pencil of rays emanating from the origin. If y_1 is set constant, say $y_1 = d$, the locus of points one gets is a circle whose Cartesian equation is (by (2))

$$x_1^2 + x_2^2 = d^2 .$$

(5)

As d varies one gets the family of polar circles. The lines (4) and circles (5) are the coordinate "lines" in the polar system which replace the families of horizontal and vertical lines in the Cartesian system. The differentials (3), which in the transformation interpretation of (2) are increments in the direction of the Cartesian axes, are in the coordinate change interpretation the first order approximations to the changes in the polar coordinates y_1 and y_2 of the point $\vec{x}$ when its Cartesian coordinates undergo the changes dx_1, dx_2.

It is reasonable to insist that any system of new coordinates supply
as refined a description of the points of the space as does the Cartesian
system. That is, it should not give the same coordinates to different
points; nor should it fail to assign well-defined coordinates to any
point. In fact this requirement is too strict to be enforced because
there are desirable curvilinear systems having points at which these
requirements are not met. Such points are called <u>degenerate</u> points. For
example, the polar system in $\mathbb{R}^2$ fails to assign well defined coordinates
at the origin, for $y_1 = 0$ there but y_2 may have any value, so the
origin is a degenerate point for the polar system. The enforceable and
customary restriction is, that a coordinate system have at most a finite
number of degenerate points. We say a coordinate system $(y_1, \ldots, y_k)$,
where $\vec{y} = \vec{f}(x)$, is <u>admissible in a region</u> E of $\mathbb{R}^k$ if at most
finitely many points in E are degenerate for the y-system. Evidently
this will be the case exactly when one can solve the equations $y_i = f^i(\vec{x})$
for $x_1, \ldots, x_k$ uniquely in terms of $y_1, \ldots, y_k$ at all but finitely
many points of E. Let the solutions be expressed as $x_i = g^i(\vec{y})$,
$i = 1, 2, \ldots, k$, or

$$\vec{x} = \vec{g}(\vec{y}) \tag{6}$$

in abused notation. For simplicity let us suppose that (6) is valid at
every point of E. Then, substituting for $\vec{y} = \vec{f}(\vec{x})$ into (6) we find
$\vec{x} = \vec{g}(\vec{f}(\vec{x}))$ for all $\vec{x}$ in E, and applying $\vec{f}$ to both sides of (6)
we find $\vec{f}(\vec{x}) = \vec{f}(\vec{g}(\vec{y}))$, which together with $\vec{y} = \vec{f}(\vec{x})$ yields
$\vec{y} = \vec{f}(\vec{g}(\vec{y}))$. In short, $\vec{y} = \vec{f}(\vec{x})$ is non-degenerate at every point of
E if there is a solution (6) valid throughout E such that

$$\vec{g}(\vec{f}(\vec{x})) \;=\; \vec{x}, \qquad \vec{f}(\vec{g}(\vec{y})) \;=\; \vec{y} \tag{7}$$

for the Cartesian and new coordinates respectively of all points of E.
If we think of $\vec{f}$ as a transformation, and write $F = \vec{f}(E)$ for the
image of the region E under the transformation, then $\vec{g}$ carries F
back onto E and (7) states that the compositions $\vec{g} \circ \vec{f}$ and $\vec{f} \circ \vec{g}$
are the <u>identity transformations</u> of E and F respectively: $\vec{g} \circ \vec{f}(\vec{x}) = \vec{x}$
for all $\vec{x}$ in E, and $\vec{f} \circ \vec{g}(\vec{y}) = \vec{y}$ for all $\vec{y}$ in F . In either
interpretation (as coordinate change or as transformation) one says in
these circumstances that $\vec{f}$ as a function is <u>invertible</u> in E, and $\vec{g}$
is its inverse function.[1] In these terms we may now state the following
useful criterion: A coordinate change $\vec{y} = \vec{f}(\vec{x})$ is admissible in any
region in which $\vec{f}$ is an invertible function.

<u>Exercise</u>. Check that the polar system is admissible in every region of $\mathbb{R}^2$
which does not contain the origin.

Suppose a given $\vec{f} : \mathbb{R}^k \to \mathbb{R}^k$ is differentiable and invertible in
a region E, and suppose that its inverse function $\vec{g}$ is also
differentiable. Then since $d(\vec{g} \circ \vec{f})(\vec{x}) = d\vec{g}(\vec{f}(\vec{x})) \cdot d\vec{f}(\vec{x})$ by 1.3(13),
and because $\vec{g} \circ \vec{f}(\vec{x}) = \vec{x}$ for all $\vec{x}$ in E, whence $d(\vec{g} \circ \vec{f})(\vec{x})$ is
the k-by-k identity matrix, we find that $d\vec{g}(\vec{f}(\vec{x}))$ and $d\vec{f}(\vec{x})$ are
mutually inverse matrices.

<u>Exercise</u>. Continuing the previous exercise, verify this for the function
$\vec{f}$ of (2) and its inverse.

Let us now interprete $\vec{f}$ as introducing new coordinates $y_i = f^i(\vec{x})$,
or (in abused notation) $\vec{y} = \vec{f}(\vec{x})$. Then $\vec{x} = \vec{g}(\vec{y})$ expresses the
Cartesian coordinates in terms of the new ones. If $d\vec{x}$ is the generic

(1) We shall have a criterion for invertibility, the so-called inverse
function theorem, in 3.1.

increment in Cartesian coordinates, we have by 1.3(12) the corresponding
increment

$$\vec{dy} = \vec{df}(\vec{x}) \cdot \vec{dx} \tag{8}$$

in the new coordinates, and since $\vec{dg}(\vec{y})$ is inverse to $\vec{df}(\vec{x})$ this
implies

$$\vec{dx} = \vec{dg}(\vec{y}) \cdot \vec{dy} . \tag{9}$$

Thus, if we were to interprete the increment $\vec{dy}$ resulting from $\vec{dx}$
as generic in the new system and inquire what increment in the old system
$\vec{dy}$ would generate, we would find it to be the original increment $\vec{dx}$
we started with. In this sense the two coordinate systems are on the
same footing. Moreover, if we were to construct the heirarchy of form
spaces based upon $(dy_1, \ldots, dy_k)$, and for this purpose it clearly
suffices to set down the one-forms

$$\alpha = \sum_j b^j(\vec{y}) \, dy_j \tag{10}$$

in the y-system, we have at once a one-to-one correspondence of the forms
(10) with the one-forms Λ^1 in Cartesian coordinates: namely, substi-
tuting for $\vec{y} = \vec{f}(\vec{x})$ and $dy_j = \sum_k f_k^j(\vec{x}) \, dx_k$ in (10) we find for α
the expression $\sum_k a^k(\vec{x}) \, dx_k$ in the Cartesian system, where $a^k(\vec{x}) =$
$\sum_j f_k^j(\vec{x}) b^j(f(\vec{x}))$; and clearly the process is reversible because of the
invertibility of $\vec{f}$. Equally clearly, this correspondence preserves the
addition of forms and their S-multiplication. We now inquire whether it

preserves the wedge product of forms. To this end it suffices to check

that the basic property 2.3(1) is preserved. That is, assuming

$dx_i \wedge dx_j = -dx_j \wedge dx_i$, does it follow that $dy_i \wedge dy_j = -dy_j \wedge dy_i$?

It does, as we now show. $dy_i \wedge dy_j = (\sum_k f_k^i dx_k) \wedge \sum_\ell f_\ell^j dx_\ell) =$

$\sum_{k<\ell} (f_k^i f_\ell^j - f_\ell^i f_k^j) dx_i \wedge dx_j$, and from this it is evident that the inter-

change of i and j will cause the right side to change sign.

We have now shown that, with regard to the algebraic properties so

far developed for differential forms, it does not matter what coordinate

system is used so long as it is related to the Cartesian system by an

invertible differentiable function with a differentiable inverse.[1] By

the same general argument we may assert even more; namely, that any two

coordinate systems which are related by an invertible differentiable

function with a differentiable inverse generate algebraically equivalent

spaces of differential forms. Thus our original choice of the Cartesian

system was merely a convenient one and not compelled.

---

(1)   The theorem of elementary calculus that the inverse of a differentiable
      function is differentiable generalizes to the present context, so that
      the final phrase above is redundant. See Courant [1], volume II, p. 152.

Chapter 3
Integration in higher dimensions

3.1 Jacobians

Let P be a rectangular parallelopiped in $\mathbb{R}^k$, with edges along the Cartesian coordinate axes. We call such a figure a Cartesian rectangle. The figure shows a Cartesian rectangle in $\mathbb{R}^2$. Let the edges of P have lengths

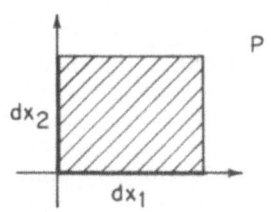

$dx_1, dx_2, \ldots, dx_k$. Then its k-dimensional volume, denoted Vol(P), is by definition

$$\text{Vol}(P) \;=\; dx_1\, dx_2\, \cdots\, dx_k \;,\tag{1}$$

the product of its edge lengths. Let T be a k-by-k matrix. We may apply T to each vector lying in P and in this way transform P to another figure T(P), which typically will no longer be a Cartesian rectangle, nor even rectangular. To calculate the volume Vol(T(P)) of T(P) we use the theorem on the geometric meaning of the determinant which states, in the present notation, that

$$\frac{\text{Vol}(T(P))}{\text{Vol}(P)} = |T|, \qquad (2)$$

where $|T|$ is the standard notation for the determinant of T.[1]
Putting this together with (1) we have

$$\text{Vol}(T(P)) = |T| dx_1 dx_2 \cdots dx_k . \qquad (3)$$

Let a differentiable function $\vec{f} : \mathbb{R}^k \to \mathbb{R}^k$ be given, and take for T the differential $d\vec{f}(\vec{x}_o)$ of $\vec{f}$ at some fixed point $\vec{x}_o$, $\vec{f}$ being interpreted as a transformation. Then by (3) we have

$$\text{Vol}(d\vec{f}(\vec{x}_o)(P)) = |d\vec{f}(\vec{x}_o)| dx_1 dx_2 \cdots dx_k . \qquad (4)$$

If P is a sufficiently small rectangle then, by the Taylor development 1.3(12), for each point $\vec{x}$ of P the image $\vec{f}(\vec{x})$ is well approximated by $d\vec{f}(\vec{x}_o) \cdot \vec{x}$, where $\vec{x}_o$ is, say, the midpoint of P. Therefore we would expect that, if P is sufficiently small, the volume $\text{Vol}\,\vec{f}(P)$ of the image of P under $\vec{f}$ should be well approximated by $\text{Vol}\,d\vec{f}(\vec{x}_o)(P)$, in some appropriate sense which we shall not inquire into here. Assuming this reasonable expectation to be so, we may write, for P sufficiently small,

$$\text{Vol}\,\vec{f}(P) \underset{\sim}{\sim} |d\vec{f}(\vec{x}_o)| dx_1 dx_2 \cdots dx_k . \qquad (5)$$

Assume now that $\vec{f}$ is invertible on a set E containing P, let $\vec{g}$

---

(1) See Courant [1], volume II, p. 33 for the theorem in dimensions 2 and 3, and Schreier-Sperner [5], p. 63 for the general case.

be the inverse in E of $\vec{f}$, and write $\vec{y}_o = \vec{f}(\vec{x}_o)$. Then (5) may be recast as

$$dx_1 \ldots dx_k \simeq |d\vec{g}(\vec{y}_o)| \text{Vol } \vec{f}(P).$$ (6)

If $F : \mathbb{R}^k \to \mathbb{R}^1$ is defined in E, then $F \circ \vec{g}$ is defined in $\vec{f}(E)$: $F \circ \vec{g}(\vec{y}) = F(\vec{x})$, where $\vec{g}(\vec{y}) = \vec{x}$, which is to say $\vec{f}(\vec{x}) = \vec{y}$. Then

$$F(\vec{x}_o) dx_1 \ldots dx_k \simeq F \circ \vec{g}(\vec{y}_o) |d\vec{g}(\vec{y}_o)| \text{Vol } \vec{f}(P).$$ (7)

If we partition E into small Cartesian rectangles, write the relation (7) for each rectangle, and add on both sides, we shall have on the left side an approximation to $\int \ldots \int_E F(x_1, \ldots, x_k) dx_1 dx_2 \ldots dx_k$, and on the right side an approximation to the integral of $F \circ \vec{g}(\vec{y}) \cdot |d\vec{g}(\vec{y})|$ over $\vec{f}(E)$. The mathematical notation for the latter would be $\int \ldots \int_{\vec{f}(E)} F \circ \vec{g}(y_1, \ldots, y_k) |d\vec{g}(\vec{y})| dy_1 dy_2 \ldots dy_k$. If all the approximations involved become exact as the number of partitioning rectangles of E increases indefinitely, we would have the following formula:

$$\int \ldots \int_E F(\vec{x}) dx_1 \ldots dx_k = \int \ldots \int_{\vec{f}(E)} F \circ \vec{g}(\vec{y}) |d\vec{g}(\vec{y})| dy_1 \ldots dy_k,$$ (8)

where $\vec{g}$ is the inverse of $\vec{f}$. We have not proved this formula; we have only argued for it on heuristic grounds. Nevertheless, having it before us, we now recognize (8) as the formula for changing the variables in integration. That is, we interprete $\vec{f}$ as a coordinate change, look upon E and $\vec{f}(E)$ as setting the limits of integration in the Cartesian and new coordinates respectively, and see the expression $|d\vec{g}(\vec{y})|\,dy_1 \cdots dy_k$

as the expression of the Cartesian volume element $dx_1 \cdots dx_k$ in terms

of the new coordinates.[1] This suggests viewing $dy_1 \cdots dy_k$ as the

y-coordinate volume element, so to speak, so that

$$dx_1 \cdots dx_k = |d\vec{g}(\vec{y})|\,dy_1 \cdots dy_k \,,$$

$$(9)$$

$$dy_1 \cdots dy_k = |d\vec{f}(\vec{x})|\,dx_1 \cdots dx_k \,,$$

where $\vec{f}$ and $\vec{g}$ are mutually inverse and $\vec{y} = \vec{f}(\vec{x})$. So far the

expression $dy_1 \cdots dy_k$ and the relations (9) are merely heuristic,

suggested by (5) and (6) together with the customary notation for integrals,

as seen in (8). We will shortly see that differential forms supply an

apparatus whereby such expressions acquire a sense which validates

relations of the type (9), and whereby formulae of the type (8) are

correctly calculated automatically. Let us first have an example for (8).

Example. Let $\iint\limits_{E} e^{-x^2-y^2}\,dx\,dy$ be required, where E is the unit disc

(boundary and interior) in $\mathbb{R}^2$. Put

$$r = f^1(x,y) = \sqrt{x^2+y^2} \,,$$

$$\theta = f^2(x,y) = \text{Arctan}\,\frac{y}{x} \,,$$

which is 2.4(8) in more familiar notation. Then $\vec{f}(E)$ specifies the

limits $0 \le r \le 1, \quad 0 \le \theta < 2\pi$; for the area element $dx\,dy$ we put the

customary (and correct) expression $r\,dr\,d\theta$; and for $F(x,y) = e^{-x^2-y^2}$

---

(1) See Courant [1], volume II, p. 247.

we put $F \circ \vec{g}(r, \theta)$, where

$$x = g^1(r, \theta) = r \cos \theta,$$

$$y = g^2(r, \theta) = r \sin \theta;$$

and we find $F \circ \vec{g}(r, \theta) = e^{-r^2}$. We find $|d\vec{g}| = r$, whence the term r

in $r\,dr\,d\theta$ is the term $|d\vec{g}|$ in the right side of (8). We have

$$\iint_E e^{-x^2-y^2} dx\,dy = \iint_{\vec{f}(E)} e^{-r^2} r\,dr\,d\theta = 2\pi \int_0^1 e^{-r^2} r\,dr = \pi(1-1/e).$$

Given a differentiable function $\vec{f}$, regarded as introducing new

coordinates $y_i = f^i(\vec{x})$, let us calculate the expression of the k-form

$dy_1 \wedge \ldots \wedge dy_k$ in Cartesian coordinates. We have $dy_1 \wedge \ldots \wedge dy_k =$

$(\sum_{i_1} f^1_{i_1} dx_{i_1}) \wedge (\sum_{i_2} f^2_{i_2} dx_{i_2}) \wedge \ldots \wedge (\sum_{i_k} f^k_{i_k} dx_{i_k})$, where each of the indices

$i_1, i_2, \ldots, i_k$ ranges over the values $1, 2, \ldots, k$. By systematic (and

unrelenting) use of the rule 2.3(1) in general form in multiplying this

expression through, we shall come out finally with

$$dy_1 \wedge \ldots \wedge dy_k = \{\sum_{i_1, \ldots, i_k} \varepsilon(i_1, \ldots, i_k) f^1_{i_1} f^2_{i_2} \ldots f^k_{i_k}\} dx_1 \wedge \ldots \wedge dx_k, \quad (10)$$

where $\varepsilon(i_1, \ldots, i_k) = \pm 1$ according as the number of interchanges

required to bring $(i_1, i_2, \ldots, i_k)$ to the order $(1, 2, \ldots, k)$ is even or

odd. This calculation may seem inscrutable at first sight. It becomes

quite simple in the case of two variables, as follows. Given $f : \mathbb{R}^2 \to \mathbb{R}^2$,

we have $dy_1 \wedge dy_2 = df^1 \wedge df^2 = (f^1_1 dx_1 + f^1_2 dx_2) \wedge (f^2_1 dx_1 + f^2_2 dx_2) =$

$f^1_1 f^2_1 dx_1 \wedge dx_1 + f^1_1 f^2_2 dx_1 \wedge dx_2 + f^1_2 f^2_1 dx_2 \wedge dx_1 + f^1_2 f^2_2 dx_2 \wedge dx_2$. The first and

last terms drop out because of the repetition; $dx_1 \wedge dx_1 = dx_2 \wedge dx_2 = 0$;

and we are left with

$$dy_1 \wedge dy_2 = (f_1^1 f_2^2 - f_2^1 f_1^2) dx_1 \wedge dx_2 , \qquad (11)$$

which has precisely the form (10). The calculation of (10) is just the one we encountered in establishing 2.3(8) but now extended to k factors. We recognize in (11) the determinant $|d\vec{f}|$ of the differential of $\vec{f}$, $|d\vec{f}| = f_1^1 f_2^2 - f_2^1 f_1^2$. It is a fact that the generalization of this expression which appears in (10) is also a determinant, being the k-by-k analog,

$$|d\vec{f}| = \sum_{i_1, \ldots, i_k} \varepsilon(i_1, \ldots, i_k) f_{i_1}^1 f_{i_2}^2 \cdots f_{i_k}^k , \qquad (12)$$

with ε as in (10). The fact that $|((f_j^i))|$ has the form (12), for any quantities f_j^i, is easily checked by hand for 3-by-3 and 4-by-4 determinants, just by regrouping their minor expansions. The proof of the general result, though not difficult, will be passed over here.[1] Putting together (12) and (10) we have

$$dy_1 \wedge \ldots \wedge dy_k = |d\vec{f}| dx_1 \wedge \ldots \wedge dx_k . \qquad (13)$$

Example. Working out (13) for the function $\vec{f} : \mathbb{R}^2 \to \mathbb{R}^2$ of 2.4(2) relating Cartesian and polar coordinates in the plane, we find

$$dy_1 \wedge dy_2 = \frac{1}{\sqrt{x_1^2 + x_2^2}} dx_1 \wedge dx_2 .$$

---

(1) The proof is given in Schreier-Sperner [5], pp. 87-89.

If we use the customary polar notation $y_1 = r$, $y_2 = \theta$, then this is

$$dr \wedge d\theta = \frac{1}{r} dx_1 \wedge dx_2 . \tag{14}$$

If the function $\vec{f}$ of (13) is invertible, then by the discussion at the end of 2.4 we shall have the companion

$$dx_1 \wedge \ldots \wedge dx_k = |d\vec{g}| dy_1 \wedge \ldots \wedge dy_k \tag{15}$$

to (13), where $\vec{g}$ is the inverse of $\vec{f}$. Moreover, for any two coordinate systems related by an invertible differentiable function with a differentiable inverse, we shall have formulae analogous to (13), (15).

The comparison of (9) with (13) and (15) leads us at once to the following decision. By the volume element in coordinates $(y_1, \ldots, y_k)$ we shall mean the k-form $dy_1 \wedge \ldots \wedge dy_k$. With this definition the heuristic relations (9) become the true formula (13), (15), and the change of variable formula (8) is now automatic: to calculate $\int \ldots \int F(\vec{x}) dx_1 \ldots dx_k$ in new variables $(y_1, \ldots, y_k)$, we first replace $dx_1 \ldots dx_k$ by $dx_1 \wedge \ldots \wedge dx_k$, substitute for the latter its expression (15) in the new variables, and of course put $F \circ \vec{g}(\vec{y})$ for $F(\vec{x})$.

Example. Continuing the previous example, from (14) we have $dx_1 \wedge dx_2 = r \, dr \, d\theta$. Thus the formula $dx_1 dx_2 = r \, dr \, d\theta$, which one would argue geometrically, becomes routine algebra in the apparatus of forms. The point of this apparatus is just that the routine algebraic method remains in situations for which the geometric intuition is obscure.

The determinant $|d\vec{f}|$ is called the Jacobian of $\vec{f}$. There are two other common notations for it: if $\vec{y} = \vec{f}(\vec{x})$ then

$$|d\vec{f}| = J(\vec{f}) = \frac{\partial(y_1, \ldots, y_k)}{\partial(x_1, \ldots, x_k)}. \tag{16}$$

The last expression is quite suggestive. If it be employed in (13), then that formula becomes the analog in k variables to the formula

$$dy = \frac{dy}{dx} dx$$

of elementary calculus.

The same notation and terminology is used in the following more general circumstance. Let $\vec{f}: \mathbb{R}^r \to \mathbb{R}^k$, $r < k$, be given. Putting $\vec{y} = \vec{f}(\vec{x})$ we have

$$y_i = f^i(x_1, \ldots, x_r), \qquad i = 1, 2, \ldots, k. \tag{17}$$

Here $d\vec{f}$ is a k-by-r matrix, so it has no determinant, but it has $\binom{k}{r}$ minor determinants of r-by-r size. These minors are called (and written as) Jacobians. Thus, if $y_{i_1}, \ldots, y_{i_r}$ is a selection of r out of the k values of i, then

$$\frac{\partial(y_{i_1}, \ldots, y_{i_r})}{\partial(x_1, \ldots, x_r)} = |((f_m^{i_\ell}))|, \tag{18}$$

where $\ell = 1, 2, \ldots, r$ and $m = 1, 2, \ldots, r$. For example, if $\vec{f}: \mathbb{R}^2 \to \mathbb{R}^3$ is given, there are $\binom{3}{2} = 3$ minors

$$\frac{\partial(y_1, y_2)}{\partial(x_1, x_2)} = f_1^1 f_2^2 - f_2^1 f_1^2 \ ,$$

$$\frac{\partial(y_1, y_3)}{\partial(x_1, x_2)} = f_1^1 f_2^3 - f_2^1 f_1^3 \ ,$$

$$\frac{\partial(y_2, y_3)}{\partial(x_1, x_2)} = f_1^2 f_2^3 - f_2^2 f_1^3 \ ,$$

corresponding to the expression of each of the three pairs of y's as functions of the two x's .

If $\vec{f} : \mathbb{R}^r \to \mathbb{R}^k$ with $r > k$ is given, we have $\binom{r}{k}$ minors of size k-by-k, the generic one being

$$\frac{\partial(y_1, \ldots, y_k)}{\partial(x_{j_1}, \ldots, x_{j_k})} = |((f_j^{i_\ell}))| \ , \tag{19}$$

where $i = 1, 2, \ldots, k$, and $\ell = 1, 2, \ldots, k$. These Jacobians measure the influence on $\vec{y}$ of the various sets $x_{j_1}, \ldots, x_{j_k}$ of k of the x_j .

Returning to k-vector-valued functions of a k-vector argument, we have the following two formulae. First, given $\vec{f}$ and $\vec{g}$, both differentiable, then $\vec{f} \circ \vec{g}$ is differentiable, as we have already noted in 1.3, and

$$J(\vec{f} \circ \vec{g}) = J(\vec{f}) \cdot J(\vec{g}) \ . \tag{20}$$

Second, if $\vec{f}$ is invertible and $\vec{f}^{-1}$ denotes its inverse, then

$$J(\vec{f}^{-1}) = J(\vec{f})^{-1} . \tag{21}$$

Both formulae are immediate consequences of the multiplication theorem for determinants. It is instructive to see these formulae in the last notation of (16). Let $\vec{u} = \vec{f}(\vec{x})$ and $\vec{v} = \vec{g}(\vec{u})$, so that $\vec{v} = \vec{f} \circ \vec{g}(\vec{x})$.

Then (20) is

$$\frac{\partial(v_1,\ldots,v_k)}{\partial(x_1,\ldots,x_k)} = \frac{\partial(v_1,\ldots,v_k)}{\partial(u_1,\ldots,u_k)} \cdot \frac{\partial(u_1,\ldots,u_k)}{\partial(x_1,\ldots,x_k)} ; \tag{22}$$

and (21) is

$$\frac{\partial(u_1,\ldots,u_k)}{\partial(x_1,\ldots,x_k)} \cdot \frac{\partial(x_1,\ldots,x_k)}{\partial(u_1,\ldots,u_k)} = 1 . \tag{23}$$

In the forms (22) and (23), both are clearly seen as generalizations to k variables of the corresponding formulae of elementary calculus; and the same for the change of variable formula (8) in this notation:

$$\int_E \cdots \int F(\vec{x}) dx_1 \wedge \ldots \wedge dx_k = \int_{\vec{f}(E)} \cdots \int F \circ \vec{g}(\vec{y}) \frac{\partial(x_1,\ldots,x_k)}{\partial(y_1,\ldots,y_k)} dy_1 \wedge \ldots \wedge dy_k .$$

One of the most basic theorems of calculus, the so-called _inverse function theorem_ mentioned above in 2.4, gives a criterion for the invertibility of functions $\vec{f} : \mathbb{R}^k \to \mathbb{R}^k$ in terms of their Jacobians. In detail the theorem asserts that, given $\vec{f} : \mathbb{R}^k \to \mathbb{R}^k$ and a point $\vec{x}_o$ at which $J(\vec{f}) \neq 0$, then there exist neighborhoods U of $\vec{x}_o$ and V of $\vec{f}(\vec{x}_o)$ such that $\vec{f}$ maps U invertibly onto V. That is, there exists a map $\vec{g}$ defined on V and taking values in U such that

$\vec{f} \circ \vec{g}(\vec{y}) = \vec{y}$ for all $\vec{y} \in V$, and $\vec{g} \circ \vec{f}(\vec{x}) = \vec{x}$ for all $\vec{x} \in U$. The theorem should be called the theorem on <u>local</u> invertibility. That is, the theorem does not imply, and it is not true, that if the Jacobian is everywhere non-zero then there is a global inverse.

<u>Example</u>. Let $\vec{f} : \mathbb{R}^2 \to \mathbb{R}^2$ be given by

$$u = f^1(\vec{x}) = e^x \sin y,$$

$$(24)$$

$$v = f^2(\vec{x}) = e^x \cos y.$$

Then $J(\vec{f}) = -e^x \sin y \cdot e^x \sin y - e^x \cos y \cdot e^x \cos y = -e^{2x} \neq 0$ for all $\vec{x}$. But the transformation (24) is not one-to-one (and so not invertible) because any two points with a vertical separation of 2π have the same image. However, it is <u>locally</u> invertible.

<u>Exercise</u>. Show that in a sufficiently small vicinity of $\vec{x} = \vec{0}$ (and state how small) one has the solution

$$x = \ln\sqrt{u^2 + v^2}$$

$$(25)$$

$$y = \text{Arctan } \frac{u}{v}$$

of (24) for x, y in terms of u, v. Note that $\vec{f}(\vec{0}) = (0, 1)$, so that near $\vec{f}(\vec{0})$ we have $v \neq 0$ and (25) is meaningful.

The inverse function theorem is a generalization to several variables of the familiar theorem of elementary calculus that if $f'(x_0) \neq 0$ for a given $f : \mathbb{R}^1 \to \mathbb{R}^1$ then f carries a neighborhood of x_0 in a one-to-one fashion onto a neighborhood of $f(x_0)$. The proof of the theorem in $\mathbb{R}^k$ is somewhat technical, and we prefer to pass over it here.[1] Let us

---

(1)   See Courant [1], volume II, p. 152.

be content with the following plausibility argument. The meaning of the Taylor development 1.3(12) is that the linear transformation $d\vec{f}(\vec{x}_o)$ approximates the transformation $\vec{f}$ in a neighborhood of $\vec{x}_o$. Now $d\vec{f}(\vec{x}_o)$ is invertible if and only if $J(\vec{f}) \neq 0$ at $\vec{x}_o$; [1] and so, if $d\vec{f}(\vec{x}_o)$ is invertible, it is plausible that $\vec{f}$ itself should be invertible near $\vec{x}_o$. This view of the theorem also serves to emphasize its local character.

3.2 Implicit Function Theorem

We denote by $Z(\phi)$ the set of zeros of $\phi : \mathbb{R}^k \to \mathbb{R}^1$. That is, $Z(\phi)$ is the set of all $\vec{x}$ such that $\phi(\vec{x}) = 0$.

Example. In $\mathbb{R}^3$, if $\phi(\vec{x}) = x_1^2 + x_2^2 + x_3^2 - 1$, then $Z(\phi)$ is the unit sphere.

The implicit function theorem asserts that if $\vec{a}$ is a point of $Z(\phi)$ such that one of the partials of ϕ, say ϕ_i, does not vanish there, $\phi_i(\vec{a}) \neq 0$, then one can solve for the ith coordinate of all points of $Z(\phi)$ near $\vec{a}$ in terms of the other variables x_j, $j \neq i$.

Example (Continued). We have $\phi_i = 2x_i$. Put $\vec{a} = (0,0,1)$. Then $\phi_3(\vec{a}) = 2 \neq 0$. In the vicinity of $\vec{a}$ (in fact on the whole northern hemisphere) we have

$$x_3 = \sqrt{1 - x_1^2 - x_2^2} = \psi(x_1, x_2) . \tag{1}$$

---

(1) by the theorem that a matrix has an inverse if and only if its determinant is not zero. The reader is doubtless familiar with this result, in the case of two variables, under the name "Cramer's Rule". See Schreier-Sperner [5], Sections 8, 9.

It is not possible to write x_1 as a function of x_2, x_3 near $\vec{a}$ on the sphere: viewed from the x_1 (or the x_2) axis the surface becomes vertical at $\vec{a}$. And $\phi_1(\vec{a}) = \phi_2(\vec{a}) = 0$. Note that $\phi(x_1, x_2, \psi(x_1, x_2)) \equiv 0$ for all x_1, x_2 at which ψ is defined (here the vicinity of $(0,0)$ in the equatorial plane).

The theorem is so named because it replaces an implicit functional relation $\phi(x_1, x_2, \ldots) = 0$ for x_1 by an explicit one $x_1 = \psi(x_1, \ldots)$. We have in general, as in the example,

$$\phi(x_1, x_2, \ldots, \psi, \ldots, x_k) \equiv 0 \tag{2}$$

wherever ψ is defined. The theorem is local in character, as was the inversion theorem on Jacobians. In fact it is a consequence of the latter, as we now show.

We are given $\phi : \mathbb{R}^k \to \mathbb{R}^1$, and a point $\vec{a}$ such that $\phi(\vec{a}) = 0$, $\phi_1(\vec{a}) \neq 0$, say. Define $\vec{f} : \mathbb{R}^k \to \mathbb{R}^k$ by

$$
\begin{aligned}
y_1 &= f^1(\vec{x}) = \phi(\vec{x}) , \\
y_2 &= f^2(\vec{x}) = x_2 , \\
&\;\;\vdots \\
y_k &= f^k(\vec{x}) = x_k .
\end{aligned}
\tag{3}
$$

Then

$$
d\vec{f} = \begin{pmatrix}
\phi_1 & \phi_2 & \cdots & \phi_k \\
0 & 1 & \cdots & 0 \\
& & \cdots & \\
0 & 0 & \cdots & 1
\end{pmatrix}
\tag{4}
$$

whence

$$J(\vec{f})(\vec{x}) = \phi_1(\vec{x}) . \tag{5}$$

Since $J(\vec{f})(\vec{a}) = \phi_1(\vec{a}) \neq 0$ we can solve for $x_1,\ldots,x_k$ in the vicinity of $\vec{x} = \vec{a}$:

$$x_1 = \psi^1(y_1,\ldots,y_k) ,$$

$$x_2 = \psi^2(y_1,\ldots,y_k) = y_2 ,$$

$$\vdots \tag{6}$$

$$x_k = \psi^k(y_1,\ldots,y_k) = y_k .$$

By (3) this is simply

$$x_1 = \psi^1(\phi(\vec{x}),x_2,\ldots,x_k) ,$$

$$\tag{7}$$

$$x_i = y_i , \qquad i = 2,3,\ldots,k .$$

But now, if $\vec{x}$ is a point of $Z(\phi)$, so that $\phi(\vec{x}) = 0$, we have

$$x_1 = \psi^1(0,x_2,\ldots,x_k) . \tag{8}$$

This expresses x_1 explicitly in terms of $x_2,\ldots,x_k$ (via ψ^1 with first argument set to 0) at every point of $Z(\phi)$ near $\vec{x} = \vec{a}$, and establishes the theorem QED.

<u>Example</u> (Continued). The $\vec{f}$ of (3) is here

$$y_1 = f^1(\vec{x}) = x_1 ,$$

$$y_2 = f^2(\vec{x}) = x_2 , \qquad (9)$$

$$y_3 = f^3(\vec{x}) = x_1^2 + x_2^2 + x_3^2 - 1 .$$

Then $J(\vec{f}) = 2x_3 = 2$ at $(0,0,1)$, and the inversion (6) is here

$$x_1 = y_1 = \psi^1(\vec{y}) ,$$

$$x_2 = y_2 = \psi^2(\vec{y}) , \qquad (10)$$

$$x_3 = \sqrt{1 - y_1^2 - y_2^2 + y_3^2} = \psi^3(\vec{y}) .$$

Now if $\vec{x}$ is a point of the sphere, so that $y_3 = 0$, we come out with

$$x_3 = \sqrt{1 - x_1^2 - x_2^2} = \psi^3(x_1, x_2, 0) , \qquad (11)$$

in accordance with the foregoing argument. And

$$\phi(x_1, x_2, \psi(x_1, x_2)) = x_1^2 + x_2^2 + (1 - x_1^2 - x_2^2) - 1 \equiv 0 .$$

Let $Z(\phi^1, \phi^2, \ldots, \phi^s)$, $s < k$, denote the set of simultaneous zeros of s given functions $\phi^i : \mathbb{R}^k \to \mathbb{R}^1$, $i = 1, 2, \ldots, s$. That is, $Z(\phi^1, \phi^2, \ldots, \phi^s)$ is the collection of all $\vec{x}$ in $\mathbb{R}^k$ such that

$$\phi^1(\vec{x}) = \phi^2(\vec{x}) = \ldots = \phi^s(\vec{x}) = 0 . \qquad (1)$$

Example. In $\mathbb{R}^3$, let

---

(1) In set-theoretic terms, $Z(\phi^1, \phi^2, \ldots, \phi^s)$ is the intersection $\cap Z(\phi^i)$ of $Z(\phi^1), Z(\phi^2), \ldots, Z(\phi^s)$.

$$\phi^1(\vec{x}) = x_1^2 + x_2^2 + x_3^2 - 1$$

$$\phi^2(\vec{x}) = x_1 + x_2 + x_3 - 1 . \tag{12}$$

Then $Z(\phi^1, \phi^2)$ is the intersection of the sphere with a plane; hence it is a circle in $\mathbb{R}^3$.

Look upon the functions ϕ^i as components of a function $\vec{\phi} : \mathbb{R}^k \to \mathbb{R}^s$, and write $y_i = \phi^i(x)$, $i = 1, 2, \ldots, s$. The matrix differential $d\vec{\phi}$ is

$$d\vec{\phi} = \begin{pmatrix} \phi_1^1 & \phi_2^1 & \cdots & \phi_k^1 \\ \phi_1^2 & \phi_2^2 & \cdots & \phi_k^2 \\ & & \vdots & \\ \phi_1^s & \phi_2^s & \cdots & \phi_k^s \end{pmatrix} , \tag{13}$$

an s-by-k matrix. It has $\binom{k}{s}$ Jacobian minors,

$$\frac{\partial(y_1, \ldots, y_s)}{\partial(x_{i_1}, \ldots, x_{i_s})} = \det((\phi_{i_\nu}^i)), \quad i_1 < i_2 < \ldots < i_s .^{(1)} \tag{14}$$

Example (Continued). The function $\vec{\phi} : \mathbb{R}^3 \to \mathbb{R}^2$ is here

$$y_1 = x_1^2 + x_2^2 + x_3^2 - 1 ,$$

$$y_2 = x_1 + x_2 + x_3 - 1 . \tag{15}$$

The $\binom{3}{2} = 3$ Jacobians are

---

(1) See 3.1(19).

$$\frac{\partial(y_1, y_2)}{\partial(x_1, x_2)} = 2(x_1 - x_2) ,$$

$$\frac{\partial(y_1, y_2)}{\partial(x_1, x_3)} = 2(x_1 - x_3) , \tag{16}$$

$$\frac{\partial(y_1, y_2)}{\partial(x_2, x_3)} = 2(x_2 - x_3) .$$

The implicit function theorem asserts, in these circumstances, that if $\vec{a}$ is a point of $Z(\phi^1, \phi^2, \ldots, \phi^s)$ at which one of the Jacobians (14) does not vanish, then one can solve near $\vec{a}$ for s of the coordinates of a point in $Z(\phi^1, \ldots, \phi^s)$ in terms of the remaining $k-s$. The previous result (stated on page 48) is clearly a special case of the present assertion, which is established in like manner to the special case, as follows.

Suppose $\phi^1(\vec{a}) = \ldots = \phi^s(\vec{a}) = 0$, and at $\vec{a}$

$$\frac{\partial(y_1, \ldots, y_s)}{\partial(x_1, \ldots, x_s)} \neq 0 . \tag{17}$$

Define a function $\vec{F} : \mathbb{R}^k \to \mathbb{R}^k$ by

$$u_1 = F^1(\vec{x}) = \phi^1(\vec{x}) ,$$

$$u_2 = F^2(\vec{x}) = \phi^2(\vec{x}) ,$$

$$\vdots$$

$$u_s = F^s(\vec{x}) = \phi^s(\vec{x}) , \tag{18}$$

$$u_{s+1} = F^{s+1}(\vec{x}) = x_{s+1} ,$$

$$\vdots$$

$$u_k = F^k(\vec{x}) = x_k .$$

Then

$$
d\vec{F} = \left(\left(\begin{array}{c} d\vec{\phi} \\ \left(\begin{array}{c} 0 \end{array}\right) \quad \left(\begin{array}{ccc} 1 & & 0 \\ & \ddots & \\ 0 & & 1 \end{array}\right) \end{array}\right)\right), \tag{19}
$$

where the upper s-by-k block is (13), and the lower-left (k-s)-by-s block has all entries 0 . Therefore $J(\vec{F})$ is the function $\dfrac{\partial(y_1,\ldots,y_s)}{\partial(x_1,\ldots,x_s)}$. Since $J(\vec{F}) \neq 0$ at $\vec{x} = \vec{a}$, we can invert $\vec{F}$ near $\vec{a}$:

$$
\begin{aligned}
x_1 &= G^1(\vec{u}) , \\
&\vdots \\
x_s &= G^s(\vec{u}) , \\
x_{s+1} &= u_{s+1} , \\
&\vdots \\
x_k &= u_k .
\end{aligned} \tag{20}
$$

By (18) the first s equations of (20) are simply

$$
\begin{aligned}
x_1 &= G^1(\phi^1(\vec{x}),\ldots,\phi^s(\vec{x}),x_{s+1},\ldots,x_k) , \\
&\vdots \\
x_s &= G^s(\phi^1(\vec{x}),\ldots,\phi^s(\vec{x}),x_{s+1},\ldots,x_k) .
\end{aligned} \tag{21}
$$

But now, if $\vec{x}$ is a point of $Z(\phi^1,\phi^2,\ldots,\phi^s)$, so that $\phi^1(\vec{x}) = \ldots = \phi^s(\vec{x}) = 0$, we have

$$x_1 = G^1(0,\ldots,0,x_{s+1},\ldots,x_k) ,$$

$$\vdots \tag{22}$$

$$x_s = G^s(0,\ldots,0,x_{s+1},\ldots,x_k) ,$$

which is the assertion QED.

We get something more here free of charge. Put $u_1 = u_2 = \ldots = u_s = 0$ in (20). Then (20) defines a map $\mathbb{R}^{k-s} \to \mathbb{R}^k$ with range in $Z(\phi^1,\phi^2,\ldots,\phi^s)$, for which at least one of the $\binom{k}{k-s}$ Jacobians [1] is non-zero; namely

$$\frac{\partial(x_{s+1},\ldots,x_k)}{\partial(u_{s+1},\ldots,u_k)} = 1 . \tag{23}$$

Example (Continued). Here the system (18) is

$$u_1 = x_1^2 + x_2^2 + x_3^2 - 1 ,$$

$$u_2 = x_1 + x_2 + x_3 - 1 , \tag{24}$$

$$u_3 = x_3 .$$

At $\vec{a} = (1,0,0)$ the first Jacobian of (16) is $2 \neq 0$. The inversion of (24) is rather untidy, involving radicals of longish expressions. But if we keep in mind that $u_1 = u_2 = 0$ on the circle $Z(\phi^1,\phi^2)$, and drop terms accordingly, we come readily to the solution (analog of (22))

$$x_1 = \frac{1 - x_3 + \sqrt{2x_3 - 3x^2 + 1}}{2} ,$$

$$\tag{25}$$

$$x_2 = \frac{1 - x_3 - \sqrt{2x_3 - 3x_3^2 + 1}}{2} ;$$

---

(1) See 3.1(18).

and one easily checks that x_1, x_2, x_3 as related by (25) do indeed fulfill the defining relations $x_1 + x_2 + x_3 - 1 = x_1^2 + x_2^2 + x_3^2 - 1 = 0$ (see (12)) of $Z(\phi^1, \phi^2)$. This is the analog of (2).

The implicit function theorem has a companion result, known as the theorem on functional dependence. Whereas in the former theorem we had s functions (implicit functional relations) $\phi^i : \mathbb{R}^k \to \mathbb{R}^1$ constituting the components of a map $\vec{\phi} : \mathbb{R}^k \to \mathbb{R}^s$, we now suppose we have k functions $\psi^j : \mathbb{R}^s \to \mathbb{R}^1$ of s arguments, which we look upon as the components of a map $\vec{\psi} : \mathbb{R}^s \to \mathbb{R}^k$. Write $x_i = \psi^i(u_1, \ldots, u_s)$, $i = 1, 2, \ldots, k$, and let the functions ψ^i be defined in some region D of $\mathbb{R}^s$. The theorem on functional dependence asserts that in the vicinity of a point $\vec{a}$ of D at which one of the Jacobians

$$\frac{\partial(x_{i_1}, x_{i_2}, \ldots, x_{i_s})}{\partial(u_1, u_2, \ldots, u_s)} , \qquad i_1 < \ldots < i_s \tag{26}$$

does not vanish, the functions ψ^i are functionally dependent: there exist k-s functions (implicit functional relations) $\phi^\ell : \mathbb{R}^k \to \mathbb{R}^1$ such that

$$\phi^\ell(\psi^1, \psi^2, \ldots, \psi^k) \equiv 0, \qquad \ell = 1, 2, \ldots, k-s \tag{27}$$

for all $\vec{u}$ near $\vec{a}$. For the proof, we suppose that at $\vec{u} = \vec{a}$ we have

$$\frac{\partial(x_1, x_2, \ldots, x_s)}{\partial(u_1, u_2, \ldots, u_s)} \neq 0. \tag{28}$$

Then we can invert the relations

$$x_i = \psi^i(\vec{u}), \quad i = 1,2,\ldots,s \tag{29}$$

near $\vec{u} = \vec{a}$:

$$u_i = \sigma^i(x_1, x_2, \ldots, x_s), \quad i = 1,2,\ldots,s, \tag{30}$$

whence

$$x_{s+j} = \psi^{s+j}(\vec{\sigma}(x_1, x_2, \ldots, x_s)), \quad j = 1,2,\ldots,k-s. \tag{31}$$

Expressed in terms of the variables u this is

$$\psi^{s+j}(\vec{u}) = \psi^{s+j}(\vec{\sigma}(\psi^1(\vec{u}), \ldots, \psi^s(\vec{u}))). \tag{32}$$

This is an identity in $u_1, u_2, \ldots, u_s$. Its functional form is

$$\psi^{s+j} = (\psi^{s+j} \circ \vec{\sigma})(\psi^1, \psi^2, \ldots, \psi^s). \tag{33}$$

That is, ψ^{s+j} is a function (namely, the composition $(\psi^{s+j} \circ \vec{\sigma})$)
of $\psi^1, \psi^2, \ldots, \psi^s$, and (33) is the set of $k-s$ functional relations
(27) QEF.

Example. Let $\vec{\psi} : \mathbb{R}^2 \to \mathbb{R}^3$ be

$$\psi^1(\vec{u}) = \text{Sin } u_1 \text{ Sin } u_2,$$

$$\psi^2(\vec{u}) = \text{Sin } u_1 \text{ Cos } u_2, \tag{34}$$

$$\psi^3(\vec{u}) = \text{Cos } u_1.$$

Then

$$u_1 = \text{Arc Sin} \sqrt{(\psi^1)^2 + (\psi^2)^2} \,,$$

$$u_2 = \text{Arc Tan}\left(\frac{\psi^1}{\psi^2}\right); \tag{35}$$

whence

$$\psi^3 = \text{Cos Arc Sin} \sqrt{(\psi^1)^2 + (\psi^2)^2} \,. \tag{36}$$

Here $\psi^3 \circ \vec{\sigma}$ is Cos Arc Sin.

Returning to the proof of the theorem, we get something more here free of charge. Replace the functions $\sigma^i : \mathbb{R}^s \to \mathbb{R}^1$ of (30) by functions $\tau^i : \mathbb{R}^k \to \mathbb{R}^1$, where

$$\tau^i(x_1, \ldots, x_k) = \sigma^i(x_1, \ldots, x_s), \quad i = 1, 2, \ldots, s \,. \tag{37}$$

That is, $\tau^i_\ell = 0$, $\ell = s+1, \ldots, k$, τ^i is independent of the last $k-s$ variables. This defines a map $\vec{\tau} : \mathbb{R}^k \to \mathbb{R}^s$ with $Z(\tau^1, \ldots, \tau^s) \subset \vec{\psi}(D)$, for which at least one of the $\binom{k}{s}$ Jacobians is non-zero; namely

$$\frac{\partial(u_1, \ldots, u_s)}{\partial(x_1, \ldots, x_s)} = \left\{ \frac{\partial(x_1, \ldots, x_s)}{\partial(u_1, \ldots, u_s)} \right\}^{-1} \neq 0 \,. \tag{38}$$

Given $\vec{\phi} : \mathbb{R}^k \to \mathbb{R}^s$, $s < k$, $d\vec{\phi}$ is a linear map $\mathbb{R}^k \to \mathbb{R}^s$, and its rank is maximal, $\text{rank}(d\vec{\phi}) = s$, if and only if

$$\sum_{i_1 < \ldots < i_s} \left\{ \frac{\partial(\phi^1, \phi^2, \ldots \phi^s)}{\partial(x_{i_1}, x_{i_2}, \ldots, x_{i_s})} \right\}^2 \neq 0 \,. \tag{39}$$

Given $\vec{\psi} : \mathbb{R}^s \to \mathbb{R}^k$, $s < k$, $d\vec{\psi}$ is a linear map $\mathbb{R}^s \to \mathbb{R}^k$, and its rank is maximal, $\mathrm{rank}(\vec{d\psi}) = s$,   if and only if

$$\sum_{i_1 < \ldots < i_s} \left\{ \frac{\partial(\psi^{i_1}, \psi^{i_2}, \ldots, \psi^{i_s})}{\partial(u_1, u_2, \ldots, u_s)} \right\}^2 \neq 0 . \tag{40}$$

The sums (39) and (40) are convenient and customary formulations of the condition that at least one of the Jacobians in question does not vanish. In these terms we may state concisely the foregoing two theorems side by side for easy comparison, it being understood that both are local results.

<u>Implicit</u> <u>Function</u> <u>Theorem</u>:   Given $\vec{\phi} : \mathbb{R}^k \to \mathbb{R}^s$, $s < k$,   if $\vec{d\phi}$ has maximal rank, then s of the variables in $\mathbb{R}^k$ depend explicitly on the remaining $k-s$, in $Z(\phi^1, \phi^2, \ldots, \phi^s)$, and there exists $\vec{\psi} : \mathbb{R}^{k-s} \to \mathbb{R}^k$ with range in $Z(\phi^1, \phi^2, \ldots, \phi^s)$, such that $\vec{d\psi}$ has maximal rank.

<u>Functional</u> <u>Dependence</u> <u>Theorem</u>:   Given $\vec{\psi} : \mathbb{R}^s \to \mathbb{R}^k$, $s < k$,   if $\vec{d\psi}$ has maximal rank, then the components of $\vec{\psi}$ satisfy $k-s$ functional relations $\phi^1, \ldots, \phi^{k-s}$ in the range of $\vec{\psi}$, $Z(\phi^1, \ldots, \phi^{k-s})$ is contained in the range of $\vec{\psi}$, and the linear map $\vec{d\phi}$ defined by these functions has maximal rank.

3.3 Manifolds

By dimension we mean number of degrees of freedom. By a j-dimensional manifold in $\mathbb{R}^k$ $(j \leq k)$ we mean a set of points in $\mathbb{R}^k$ each of which is determined by the specification of exactly j numbers. That statement is both too vague and too general, but let it stand as a first approximation which we will clarify in a moment. It will turn

out (after the clarification) that a j-dimensional manifold looks, in

the vicinity of each of its points, very like a small piece of $\mathbb{R}^j$.

Let $\quad \phi : \mathbb{R}^k \to \mathbb{R}^1$ be such that

$$\sum_{i=1}^{k} \{\phi_i\}^2 \neq 0 . \tag{1}$$

We are in the circumstances of the implicit function theorem: we have a

map from $\mathbb{R}^k$ to $\mathbb{R}^1$, namely ϕ itself; and (1) is the Jacobian

condition 3.2(39) for this case. Therefore we can solve locally in

$Z(\phi)$ for one of the variables, say

$$x_1 = \sigma(x_2, x_3, \ldots, x_k) . \tag{2}$$

This may not hold throughout $Z(\phi)$: we may require different sets of

k-1 x's as independent variables in different regions of $Z(\phi)$.

Example. In $\mathbb{R}^3$, with $\vec{\phi}(x) = x_1^2 + x_2^2 + x_3^2 - 1$, we had in the vicinity

of the north pole[1]

$$x_3 = \sqrt{1 - x_1^2 - x_2^2} ,$$

but this does not do in the vicinity of (1,0,0). There we must write

$$x_1 = \sqrt{1 - x_2^2 - x_3^2} .$$

---

(1) See 3.2(11)

Nevertheless, locally we may express $Z(\phi)$ in terms of $k-1$ parameters:

$$x_1 = \sigma(u_2, u_3, \ldots, u_k),$$

$$x_2 = u_2,$$

$$\vdots \tag{3}$$

$$x_k = u_k.$$

That is, the parameters $u_2, \ldots, u_k$ are just $x_2, \ldots, x_k$. We look upon (3) as a map $\vec{\psi} : \mathbb{R}^{k-1} \to \mathbb{R}^k$. That is,

$$x_1 = \psi^1(u_2, \ldots, u_k) = \sigma(u_2, \ldots, u_k),$$

$$x_2 = \psi^2(u_2, \ldots, u_k) = u_2,$$

$$\vdots \tag{4}$$

$$x_k = \psi^k(u_2, \ldots, u_k) = u_k.$$

Then

$$\frac{\partial(\psi^2, \psi^3, \ldots, \psi^k)}{\partial(x_2, x_3, \ldots, x_k)} = 1, \tag{5}$$

and in particular the Jacobian condition 3.2(40) is satisfied by the map $\vec{\psi}$.

We have shown that if $\phi : \mathbb{R}^k \to \mathbb{R}^1$ satisfies (1), then $Z(\phi)$ is a $(k-1)$-dimensional manifold, according to our provisional definition; moreover, we have shown that the parametric presentation (which demonstrates the dimension) satisfies the Jacobian condition 3.2(40) when construed as a map $\mathbb{R}^{k-1} \to \mathbb{R}^k$. For particular manifolds $Z(\phi)$ it is

often possible (and always desirable) to find parameters (other than
a subset of the x_i) which suit the geometry of the manifold and which
provide a parametric presentation valid for the whole manifold.

Example (Continued). For the sphere in $\mathbb{R}^3$ we may employ the latitude
and longitude angles u_1, u_2. Then $Z(\phi)$ is given by

$$x_1 = \text{Sin } u_1 \text{ Sin } u_2$$

$$x_2 = \text{Sin } u_1 \text{ Cos } u_2 \quad \left\{ \begin{array}{c} 0 \le u_1 \le \pi \\ 0 \le u_2 \le 2\pi \end{array} \right\} \tag{6}$$

$$x_3 = \text{Cos } u_1 .$$

Viewing (6) as a map $\vec{\psi} : \mathbb{R}^2 \to \mathbb{R}^3$ we find for the three Jacobians

$$\frac{\partial(\psi^1, \psi^2)}{\partial(u_1, u_2)} = \text{Sin } u_1 \text{ Cos } u_1 ,$$

$$\frac{\partial(\psi^1, \psi^3)}{\partial(u_1, u_2)} = \text{Sin}^2 u_1 \text{ Cos } u_2 , \tag{7}$$

$$\frac{\partial(\psi^2, \psi^3)}{\partial(u_1, u_2)} = \text{Sin}^2 u_1 \text{ Sin } u_2 ,$$

and, looking to 3.2(40), we find

$$\sum_{i<j} \left\{ \frac{\partial(\psi^i, \psi^j)}{\partial(u_1, u_2)} \right\}^2 = \text{Sin}^2 u_1 , \tag{8}$$

whence the Jacobian condition is satisfied at all points except the
north and south poles.

Let $\vec{\phi}:\mathbb{R}^k \to \mathbb{R}^s$ $(s < k)$ fulfill the Jacobian condition 3.2(39).
Then, by a discussion following the steps we have taken above for a
single function $\phi:\mathbb{R}^k \to \mathbb{R}^1$, we find that $Z(\phi^1,\phi^2,\ldots,\phi^s)$ is a
manifold of dimension $k-s$, and that the parametric presentation which
demonstrates the dimension fulfills 3.2(40) when construed as a map
$\mathbb{R}^{k-s} \to \mathbb{R}^k$. For particular manifolds it is often possible to find a
parametrization valid over the whole manifold and satisfying 3.2(40)
at all but at most a finite number of points (as in the example). With
this as background, we now state the

Definition. A j-dimensional manifold in $\mathbb{R}^k$, $j \le k$, is the image under
a map $\vec{\psi}:\mathbb{R}^j \to \mathbb{R}^k$ of a rectangular region D in R^j, such that at
all but at most a finite number of points of D the Jacobian condition
3.2(40) is fulfilled. By a rectangular region in $\mathbb{R}^j$ we mean a
j-dimensional rectangular parallelopiped which may have infinite extent
in any of the coordinate directions.

A set of the form $Z(\phi^1,\phi^2,\ldots,\phi^\ell)$ for which the appropriate
differential has maximal rank is called a _variety_. The content of the
two theorems of 3.2 is: a variety is locally a manifold, and a manifold
is locally a variety.

Let $M = \vec{\psi}(D)$, $\vec{\psi}:\mathbb{R}^r \to \mathbb{R}^s$, $r < k$, be a manifold of dimension r.
The components ψ^i give rise by partial differentiation to r vectors
$\vec{\psi}_\nu$ in $\mathbb{R}^k$:

$$\vec{\psi}_\nu = (\psi^1_\nu, \psi^2_\nu, \ldots, \psi^k_\nu), \qquad \nu = 1, 2, \ldots, r \ . \tag{9}$$

For any r scalars $\vec{a} = (a_1, a_2, \ldots, a_r)$ we have

$$(\vec{d\psi})\vec{a} \;=\; \textstyle\sum a_\nu \vec{\psi}_\nu \;. \tag{10}$$

Since M is a manifold, the Jacobian condition 3.2(40) is in force for $\vec{\psi}$, which is to say, $\vec{d\psi}$ has rank r . Therefore if $\sum a_\nu \vec{\psi}_\nu = 0$, then $\vec{a} = \vec{0}$, $a_1 = a_2 = \ldots = a_r = 0$, by (10). That is, the vectors (9) are linearly independent.

M is locally a variety $Z(\phi^1,\phi^2,\ldots,\phi^s)$, $\phi^\mu : \mathbb{R}^k \to \mathbb{R}^1$, $s+r = k$. By partial differentiation the functions ϕ^μ give rise to s vectors $\vec{\phi}^\mu$ in $\mathbb{R}^k$:

$$\vec{\phi}^\mu \;=\; (\phi^\mu_1, \phi^\mu_2, \ldots, \phi^\mu_k), \qquad \mu = 1, 2, \ldots, s \;. \tag{11}$$

For any s scalars $\vec{b} = (b_1, b_2, \ldots, b_s)$ we have[1]

$$^t(\vec{d\phi})\vec{b} \;=\; \textstyle\sum b_\mu \vec{\phi}^\mu \;. \tag{12}$$

By the Jacobian condition 3.2(30), $^t(\vec{d\phi})$ has rank s . Therefore if $\sum b_\mu \vec{\phi}^\mu = 0$, then $\vec{b} = \vec{0}$, $b_1 = \ldots = b_s = 0$, by (12). That is, the vectors (11) are linearly independent.

Because M is locally $Z(\phi^1,\phi^2,\ldots,\phi^s)$, we have

$$\phi^\mu(\psi^1, \ldots, \psi^2) \;\equiv\; 0 \;. \tag{13}$$

Differentiating through (13) by the νth argument of $\vec{\psi}$ we find

---

(1) ^tA is the transpose of the matrix A .

$$\sum_{i=1}^{k} \phi_i^\mu \psi_\nu^i = 0, \tag{14}$$

which is to say

$$\vec{\phi}^\mu \cdot \vec{\psi}_\nu = 0, \quad \mu = 1, \ldots, s; \quad \nu = 1, \ldots, r, \tag{15}$$

each of the vectors (9) is orthogonal to each of the vectors (11).

Look upon the vectors (9) as attached to a point $\vec{x}$ of M. They form a space of dimension r, which is to say $\mathbb{R}^r$, which touches M at $\vec{x}$. If the components $\delta_1, \delta_2, \ldots, \delta_r$ of a vector $\vec{\delta}$ are sufficiently small, then by (10) we have

$$\psi(u + \vec{\delta}) - \psi(\vec{u}) \stackrel{\sim}{=} \sum \delta_\nu \vec{\psi}_\nu, \tag{16}$$

where $\vec{\psi}(\vec{u}) = \vec{x}$, and $\stackrel{\sim}{=}$ denotes linear approximation. This means that all the vectors (9) are tangent to M at $\vec{x}$. A linear space with this property is called (appropriately) a tangent space to M. By (15) there can be no vectors other than the span of (9) tangent to M, and we have shown that the tangent space to M (space spanned by all vectors tangent to M) is $\mathbb{R}^r$. In this sense a manifold of dimension r is, in the vicinity of each of its points, very like a small piece of $\mathbb{R}^r$.

We have also of course shown that the normal space to M (space spanned by all vectors orthogonal to M) is $\mathbb{R}^s$, $r + s = k$.

Example. In $\mathbb{R}^2$ the unit circle S^1 is $Z(\phi)$ where $\phi = x_1^2 + x_2^2 - 1$. The tangent space at each point of S^1 is the line (i.e., $\mathbb{R}^1$) tangent

to the circle; the normal space is
the line ($\mathbb{R}^1$)   perpendicular to
it at the point (see Figure).   There
is no use trying to draw higher
dimensional figures, but the analogy
with this case is exact.

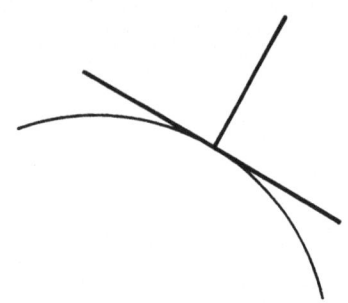

If   M   is a 2-dimensional manifold (surface) in $\mathbb{R}^3$,   given say
by $\vec{\psi} : \mathbb{R}^2 \to \mathbb{R}^3$,   the tangent space is spanned by $\vec{\psi}_1, \vec{\psi}_2$,   and the
normal space, which is 1-dimensional, is neatly described as the span
of the vector product $\vec{\psi}_1 \times \vec{\psi}_2$.

<u>Example</u>. For the sphere S^2 in $\mathbb{R}^3$ as given in terms of the latitude
and longitude angles (6), we have

$$\vec{\psi}_1 = (\text{Cos } u_1 \text{ Sin } u_2, \text{ Cos } u_1 \text{ Cos } u_2, - \text{Sin } u_1) ,$$

$$\vec{\psi}_2 = (\text{Sin } u_1 \text{ Cos } u_2, - \text{Sin } u_1 \text{ Sin } u_2, 0) . \tag{17}$$

The calculation for $\vec{\psi}_1 \times \vec{\psi}_2$ has already been done in (7).   The result
is

$$\vec{\psi}_1 \times \vec{\psi}_2 = (-\text{Sin}^2 u_1 \text{ Sin } u_2, - \text{Sin}^2 u_1 \text{ Cos } u_2, - \text{Sin } u_1 \text{ Cos } u_1) , \tag{18}$$

and notice that $\| \vec{\psi}_1 \times \vec{\psi}_2 \|^2$ is exactly the left side of the Jacobian
condition 3.2(40) for this case.   That condition thus asserts that there
is a non-zero normal vector at every point of S^2,   and, since the
vector in question is $\vec{\psi}_1 \times \vec{\psi}_2$,   this says the tangent space is

2-dimensional at every point of M (or, in a word, is non-degenerate). Since $S^2 = Z(\phi)$, with $\phi = x_1^2 + x_2^2 + x_3^2 - 1$, the normal space is also spanned by

$$\vec{\phi} = (2x_1, 2x_2, 2x_3) ; \tag{19}$$

and by (6) we find

$$\vec{\psi}_1 \times \vec{\psi}_2 = (-\tfrac{1}{2} \operatorname{Sin} u_1) \vec{\phi} . \tag{20}$$

Our discussion in 2.1 of curves in $\mathbb{R}^k$ falls within the purview of this section. We had a map $\vec{\gamma} : \mathbb{R}^1 \to \mathbb{R}^k$; the tangent space is 1-dimensional, spanned by $\vec{T} = (\tfrac{d}{dt} \gamma^1, \ldots, \tfrac{d}{dt} \gamma^k)$, which is (9) for this case; and the Jacobian condition 3.2(40) is here the requirement that $\|\vec{T}\| \neq 0$ at every point of the curve (or, that the tangent space be everywhere non-degenerate).

3.4 Integration on Manifolds

Given a manifold M of dimension r in $\mathbb{R}^k$, $r < k$, we want to find the r-dimensional volume element on M. We have already done so for curves in $\mathbb{R}^k$; the answer is the line-element formula 2.1(5). One might expect, after our work in 3.1 on Jacobians, that the volume element we seek should be a Jacobian. It cannot be directly so expressed in terms of the parameters of M, for one has k coordinates expressed in terms of $r < k$ parameters. One can express the volume element in

terms of the parameters by appealing to the geometry of the tangent
space. More precisely, we make use of the disposition within the tangent
space of the basis 3.3(9); and the volume element will be determined by
the r-dimensional volume of the r-dimensional parallelopiped in the
tangent space determined by the basis 3.3(9). For a curve, this volume
is merely the length of the tangent vector, and so 2.1(5) is an instance
of the procedure. The procedure is justified by 3.3(16): if the
parameters undergo small changes, then the changes of the corresponding
point of M in the r independent directions are given to first (linear)
approximation by the changes in the basic tangent directions 3.3(9);
and so the volume of the curvilinear "parallelopiped" in M so generated
is to first approximation that of the corresponding parallelopiped in the
tangent space.

Let M be a 2-dimensional manifold (a surface) in $\mathbb{R}^k$, given, say,
by $\vec{\psi}: \mathbb{R}^2 \to \mathbb{R}^k$. The two basic tangent vectors are

$$\vec{\psi}_\ell = (\psi_\ell^1, \ldots, \psi_\ell^k), \quad \ell = 1, 2, \tag{1}$$

and the area A of the parallelogram they generate in the tangent plane
is readily calculated by the law of cosines. We have

$$A^2 = \|\vec{\psi}_1\|^2 \|\vec{\psi}_2\|^2 \left\{ 1 - \frac{(\vec{\psi}_1 \cdot \vec{\psi}_2)^2}{\|\vec{\psi}_1\|^2 \|\vec{\psi}_2\|^2} \right\}$$

$$= \sum_{i=1}^{k} (\psi_1^i)^2 \sum_{j=1}^{k} (\psi_2^j)^2 - \left(\sum_{\ell=1}^{k} \psi_1^\ell \psi_2^\ell \right)^2. \tag{2}$$

We have only to notice what cancellations accrue from the minus sign to see at once that

$$A^2 = \sum_{i<j} (\psi_1^i \psi_2^j - \psi_1^j \psi_2^i)^2 . \tag{3}$$

That is,

$$A^2 = \sum_{i<j} \left\{ \frac{\partial(\psi^i, \psi^j)}{\partial(u_1, u_2)} \right\}^2 , \tag{4}$$

which is the left side of the Jacobian condition 3.2(40) for a surface. Write dv for the volume element in general (line element for a curve, area element for a surface, etc.). By (4) we have for a surface M in $\mathbb{R}^k$

$$dv = \left(\sum_{i<j} \left\{ \frac{\partial(\psi^i, \psi^j)}{\partial(u_1, u_2)} \right\}^2 \right)^{\frac{1}{2}} du_1 \wedge du_2 . \tag{5}$$

This volume element is oriented: there are two possible orderings of the basic tangent vectors (1), or equivalently, of the parameters u_1, u_2; if the ordering is changed, the volume element changes sign.

Let M be a manifold of dimension r in $\mathbb{R}^k$, r<k, given, say, by $\vec{\psi} : \mathbb{R}^r \to \mathbb{R}^k$. By the r-dimensional analog of the preceding calculation[1] we find

$$dv = \left(\sum_{i_1 < \ldots < i_r} \left\{ \frac{\partial(\psi^{i_1}, \psi^{i_2}, \ldots, \psi^{i_r})}{\partial(u_1, u_2, \ldots, u_r)} \right\}^2 \right)^{\frac{1}{2}} du_1 \wedge \ldots \wedge du_r . \tag{6}$$

---

(1) This will be discussed in the Appendix.

This volume element is oriented: there are r! possible orderings of the parameters; the passage between two orderings is an odd or even permutation (i.e., requires an odd or even number of interchanges) of 1,2,...,r; and the sign of (6) is changed or unchanged accordingly.

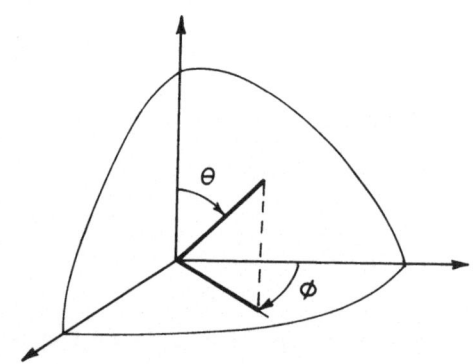

Exercise. Let θ, ϕ denote latitude and longitude respectively on the unit sphere in $\mathbb{R}^3$. Then

$$x_1 = \text{Sin } \theta \text{ Cos } \phi = \psi^1(\theta,\phi)$$

$$x_2 = \text{Sin } \theta \text{ Sin } \phi = \psi^2(\theta,\phi)$$

$$x_3 = \text{Cos } \theta \qquad = \psi^3(\theta,\phi)$$

describes the sphere. Verify that (6) yields the familiar surface area element

$$dv = \text{Sin } \theta \, d\theta \, d\phi \, .$$

We look upon the volume element as constructed out of a vector-like object $\vec{v}$ having (in the case of an r-dimensional manifold M) the $\binom{k}{r}$ components

$$v_{i_1,i_2,\dots,i_r} = \frac{\partial(\psi^{i_1},\psi^{i_2},\dots,\psi^{i_r})}{\partial(u_1,u_2,\dots,u_r)}, \qquad i_1 < \dots < i_r \, . \tag{7}$$

Then

$$dv = \|\vec{v}\| du_1 \wedge \dots \wedge du_2 \, , \tag{8}$$

where $\|\vec{v}\|$ is the $\binom{k}{r}$-dimensional length of $\vec{v}$. For a curve, $\vec{v}$

has $\binom{k}{1} = k$ components and is an ordinary vector, $\vec{v} = \vec{T}$; whence

the line element formula 2.1(5) is a special case of (8). We look upon

an r-form $\omega = \sum_{i_1 < \ldots < i_r} a^{i_1, i_2, \ldots, i_r} dx_{i_1} \wedge \ldots \wedge dx_{i_r}$ as a vector-like

object $\vec{\omega}$ having the $\binom{k}{r}$ components

$$\omega_{i_1, i_2, \ldots, i_r} = a^{i_1, i_2, \ldots, i_r}, \qquad i_1 < \ldots < i_r . \tag{9}$$

As $\vec{x}$ ranges over a manifold M of dimension r, the components (9)
are then functions of the parameters $\vec{u}$, $a^{i_1, \ldots, i_r}(\vec{x}) = a^{i_1, \ldots, i_r}(\psi(\vec{u}))$
The dot product $\vec{\omega} \cdot \vec{v}$ (which is a $\binom{k}{r}$-dimensional dot product) is
then a scalar-valued function defined on the region D in R^r which
specifies through ψ the manifold M. The integral over D of this
scalar function is one definition of the integral of an r-form over an
r-manifold:

$$\int_M \omega = \int \ldots \int_D \vec{\omega} \cdot \vec{v} \, du_1 \wedge \ldots \wedge du_r . \tag{10}$$

This is the r-dimensional analog of the integral 2.1(7) of a one-form
(vector field) over a curve. That formula was interpreted as the integral
over the curve of the component along the tangent of the vector field.
So too here: the integral (10) is the component of $\vec{\omega}$ along $\vec{v}$
integrated over M. That component is $\vec{\omega} \cdot \dfrac{\vec{v}}{\|\vec{v}\|}$, so that the integral

just described would be $\int \vec{\omega} \cdot \dfrac{\vec{v}}{\|\vec{v}\|} \, dv = \int \vec{\omega} \cdot \dfrac{\vec{v}}{\|\vec{v}\|} \|\vec{v}\| du_1 \wedge \ldots \wedge du_r$, which is (10). Passing to a more abstract notation, analogous to 2.1(8), we write

$$\int_M \omega = \int_M \left\{ \sum_{i_1 < \ldots < i_r} a^{i_1, i_2, \ldots, i_r} dx_{i_1} \wedge \ldots \wedge dx_{i_r} \right\}. \tag{11}$$

This integral transforms itself into (10) when a parametrization $\vec{\psi}$ of M is introduced; for then, $dx_i = \sum_j \psi_j^i du_j$, and

$$dx_{i_1} \wedge \ldots \wedge dx_{i_r} = \frac{\partial(\psi^{i_1}, \ldots, \psi^{i_r})}{\partial(u_1, \ldots, u_r)} du_1 \wedge \ldots \wedge du_r$$

$$\tag{12}$$

$$= v_{i_1, \ldots, i_r} du_1 \wedge \ldots \wedge du_r .$$

This abstract formulation is not meaningful (indeed, the integral (10) is not well defined) unless we can show that the integral is invariant under change of coordinates. To this end, suppose that M is also expressible in terms of new coordinates $\vec{v} = (v_1, \ldots, v_r)$. Assuming the u's and v's are both admissible coordinate systems, there is then a map $\vec{\phi} : \mathbb{R}^r \to \mathbb{R}^r$, $\vec{u} = \vec{\phi}(\vec{v})$, such that $J(\vec{\phi}) \neq 0$ at every point of $(\vec{\phi})^{-1}(D)$. The components of the form ω are simply expressed in terms of the new coordinates by functional composition:

$$a^{i_1, \ldots, i_r}(\vec{\psi}(\vec{u})) = a^{i_1, \ldots, i_r}(\vec{\psi} \circ \vec{\phi}(\vec{v})) .$$

By 3.1(13) we have

$$dv_1 \wedge \ldots \wedge dv_r = \frac{\partial(v_1, \ldots, v_r)}{\partial(u_1, \ldots, u_r)} du_1 \wedge \ldots \wedge du_r .$$

For any fixed indices $\quad i_1 < \ldots < i_r \quad$ the map with components
$\psi^{i_1} \circ \vec{\phi}, \; \psi^{i_2} \circ \vec{\phi}, \; \ldots, \; \psi^{i_r} \circ \vec{\phi} \quad$ is the composition of two maps $\quad \mathbb{R}^r \to \mathbb{R}^r$.

Therefore, by the multiplication theorem for determinants (see 3.1(20)),

$$\frac{\partial(\psi^{i_1}, \ldots, \psi^{i_r})}{\partial(u_1, \ldots, u_r)} \cdot \frac{\partial(u_1, \ldots, u_r)}{\partial(v_1, \ldots, v_r)} = \frac{\partial(\psi^{i_1}, \ldots, \psi^{i_r})}{\partial(v_1, \ldots, v_r)} \; ;$$

and so

$$\int_D \ldots \int \sum a^{i_1, \ldots, i_r}(\vec{\psi}(\vec{u})) \frac{\partial(\psi^{i_1}, \ldots, \psi^{i_r})}{\partial(u_1, \ldots, u_r)} \, du_1 \wedge \ldots \wedge du_r =$$

$$\int_{(\vec{\phi})^{-1}(D)} \ldots \int \sum a^{i_1, \ldots, i_r}(\vec{\psi} \circ \vec{\phi}(\vec{v})) \frac{\partial(\psi^{i_1}, \ldots, \psi^{i_r})}{\partial(v_1, \ldots, v_r)} \frac{\partial(v_1, \ldots, v_r)}{\partial(u_1, \ldots, u_r)} \frac{\partial(u_1, \ldots, u_r)}{\partial(v_1, \ldots, v_r)} \, dv_1 \wedge \ldots \wedge dv_r$$

We observe that $\quad J(\vec{\phi}) = \dfrac{\partial(u_1, \ldots, u_r)}{\partial(v_1, \ldots, v_r)} \quad$ appears accompanied by its

reciprocal, so that it cancels, and consequently $\int_M \omega$ is the same in

any coordinate system.

The integral (11) of an r-form over an r-manifold is a generalization

to r dimensions of the integral 2.1(8) of a vector field over a curve.

In this sense, r-forms are to r-dimensional surface integrals as one-forms

are to line integrals.

Chapter 4
Exterior differentiation

4.1 Exterior Derivative

The _exterior derivative_ is an operation on forms which extends to higher degrees the process by which the one-form df is got from a zero-form (scalar field) f ; and the same notation (prefix d) is used. The definition has two parts.

First, if $a(\vec{x})dx_{i_1} \wedge \ldots \wedge dx_{i_r}$, $i_1 < \ldots < i_r$, is a single component of an r-form in $\mathbb{R}^k$, $r < k$, then

$$d(a \; dx_{i_1} \wedge \ldots \wedge dx_{i_r}) \; = \; (da) \wedge dx_{i_1} \wedge \ldots \wedge dx_{i_r} \tag{1}$$

by definition. This expression may be evaluated by introducing a new subscript i_0 . Then

$$d(a \; dx_{i_1} \wedge \ldots \wedge dx_{i_r}) \; =$$

$$\tag{2}$$

$$\sum_{i_0=1,2,\ldots,k; \; i_\nu < i_0 < i_{\nu+1}} (-1)^\nu a_{i_\nu} \; dx_{i_1} \wedge \ldots \wedge dx_{i_\nu} \wedge dx_{i_0} \wedge dx_{i_{r+1}} \wedge \ldots \wedge dx_{i_r} \; .$$

We have taken the ordinary differential $da = \sum_i a_i dx_i$ of a, which we write as $\sum_{i_0} a_{i_0} dx_{i_0}$, and then in taking the wedge product ((1),

right side) we have to adjust signs according to the number of trans-

positions required to put the (r+1) basic one-forms $dx_{i_0}, dx_{i_1}, \ldots, dx_{i_r}$

in proper order. The strict inequalities $i_\nu < i_0 < i_{\nu+1}$ accord with the

rule 2.3(2): terms with equal indices are zero. Note that the result

(2) is an (r+1)-form.

The definition of the exterior derivative is completed by the

requirement that if ω, τ are r-forms and a, b are numbers (constants)

then

$$d(a\omega + b\tau) = a\, d\omega + b\, d\tau \; ; \tag{3}$$

that is, d is a linear operation in the ordinary sense of that term.

Putting (2) and (3) together we have the following formula for the

exterior derivative of an r-form:

$$d\left(\sum_{i_1 < \ldots < i_r} a^{i_1, \ldots, i_r} dx_{i_1} \wedge \ldots \wedge dx_{i_r} \right) =$$

$$\sum_{i_0 < i_1 < \ldots < i_r} \left(\sum_{\nu=0}^{r} (-1)^\nu a^{i_0, \ldots, \hat{i}_\nu, \ldots, i_r} \right) dx_{i_0} \wedge dx_{i_\nu} \wedge \ldots \wedge dx_{i_r} . \tag{4}$$

The notation $a^{i_0, i_1, \ldots, \hat{i}_\nu, \ldots, i_r}$ means that the superscript i_ν

is deleted.

Formula (4) is cumbersome. We have written it here only to show

that there is a general formula. In practice one calculates $d\omega$ from

(1), (3), and the rules 2.3(1) and 2.3(2). Here are a few examples in low degrees.

Example 1. $df = \sum_i f_i dx_i$.

Example 2. $d(\sum_j a^j dx_j) = \sum_j (da^j) \wedge dx_j = \sum_j (\sum_i a^j_i dx_i) \wedge dx_j = \sum_{i<j} (a^j_i - a^i_j) dx_i \wedge dx_j$.

Example 3. $d(\sum_{i<j} a^{ij} dx_i \wedge dx_j) = \sum_{i<j} d(a^{ij}) \wedge dx_i \wedge dx_j =$

$\sum_{i<j} (\sum_\ell a^{ij}_\ell dx_\ell) \wedge dx_i \wedge dx_j = \sum_{\ell<i<j} (a^{ij}_\ell - a^{\ell j}_i + a^{\ell i}_j) dx_\ell \wedge dx_i \wedge dx_j$.

Forms of top degree have exterior derivative 0:

$$\omega \text{ in } \wedge^k \implies d\omega = 0 .\tag{5}$$

For by (2), or (4), $d\omega$ will involve a wedge product of (k+1) of the basic one-forms $dx_1, \ldots, dx_k$, whence there must be a repetition.

We summarize: the exterior derivative d as defined by (1) and (3) is a linear map $d : \wedge \to \wedge$ such that

$$d : \wedge^r \to \wedge^{r+1}, \quad r = 0, 1, \ldots, k-1 ; \quad d : \wedge^k = 0 .\tag{6}$$

As a consequence of the Leibniz Rule for the ordinary differentiation of ordinary products we have for the exterior derivative of wedge products of forms the rule

$$d(\omega \wedge \tau) = (d\omega) \wedge \tau + (-1)^{\deg \omega} \tau \wedge d\omega .\tag{7}$$

To establish (7) it is sufficient, by (3), to consider forms with one component

$$\omega = a \; dx_{i_1} \wedge \ldots \wedge dx_{i_r} \; , \qquad \tau = b \; dx_{j_1} \wedge \ldots \wedge dx_{j_s} \qquad (8)$$

of degrees, say, r and s. If the index sets $\{i_\mu\}$ and $\{j_\nu\}$ overlap then all three terms in (7) vanish and (7) is trivially true. If, on the other hand, the index sets are disjoint, then

$$\omega \wedge \tau = ab \; dx_{i_1} \wedge \ldots \wedge dx_{i_r} \wedge dx_{j_1} \wedge \ldots \wedge dx_{j_s} \; . \qquad (9)$$

For present purposes we need not arrange the subscripts in increasing order; we take them as they come. In the same spirit

$$d(\omega \wedge \tau) = \sum_i (a_i b + a b_i) dx_i \wedge dx_{i_1} \wedge \ldots \wedge dx_{i_r} \wedge dx_{j_1} \wedge \ldots \wedge dx_{j_s} \; . \qquad (10)$$

Distributing across the plus sign, we have two groups of terms. The first is $\sum_i a_i b \; dx_i \wedge dx_{i_1} \wedge \ldots \wedge dx_{j_s} = (\sum_i a_i dx_i \wedge dx_{i_1} \wedge \ldots \wedge dx_{i_r}) \wedge$ $b \; dx_{j_1} \wedge \ldots \wedge dx_{j_s} = (d\omega) \wedge \tau$. To identify the second group in terms of given data (that is: ω, τ, $d\omega$, or $d\tau$) we shall have to pass dx_i across $dx_{i_1} \wedge \ldots \wedge dx_{i_r}$, which requires r transpositions. Adjusting the sign accordingly, this group is $\sum_i ab_i dx_i \wedge dx_{i_1} \wedge \ldots \wedge dx_{j_s} =$ $(-1)^r (a \; dx_{i_1} \wedge \ldots \wedge dx_{i_r}) \wedge (\sum_i b_i dx_i \wedge dx_{j_1} \wedge \ldots \wedge dx_{j_s}) = (-1)^r \omega \wedge d\tau$, and (7) follows, QED.

The argument makes clear why the sign in (7) is determined by the degree of the first factor alone.

Example. $d((\sum_j a^j dx_j) \wedge f) = d(\sum_j (a^j f) dx_j) = \sum_j (\sum_i (a^j f)_i dx_i) \wedge dx_j =$
$\sum_j (\sum_i (a_i^j f + a^j f_i) dx_i) \wedge dx_j =$

$$\sum_{i<j} \{(a_i^j f + a^j f_i) - (a_j^i f + a^i f_j)\} dx_i \wedge dx_j .$$ (11)

Write $\omega = \sum_j a^j dx_j$. The expression (11) can be identified (in terms of ω, f, dω, df) in two ways. On one hand, factoring the a's on the left, it is $d\omega \wedge f - \omega \wedge (df)$; on the other, factoring the a's on the right, it is $f \wedge d\omega + (df) \wedge d\omega$. Both these expressions equal $d(\omega \wedge f)$ because (11) does. By the commutation rule 2.3(18), $\omega \wedge f = f \wedge \omega$, $(d\omega) \wedge f = f \wedge (d\omega)$, and $(df) \wedge \omega = -\omega \wedge (df)$. Thus

$$d(\omega \wedge f) = (d\omega) \wedge f - \omega \wedge (df),$$

$$d(f \wedge \omega) = (df) \wedge \omega + f \wedge (d\omega),$$

in accordance with (7).

The Leibniz rule might be surmised as follows. On account of the appearance of algebraic signs in wedge and exterior derivative calculations, one would be led to assume a rule of the form

$$d(\omega \wedge \tau) = (d\omega) \wedge \tau + \varepsilon \omega \wedge (d\tau), \quad \varepsilon = \pm 1 .$$

From this assumption it follows, by use of the commutation rule, that $\varepsilon = (-1)^{\deg \omega}$. We omit the details.

The general form of Leibniz's rule is

$$d(\omega_1 \wedge \omega_2 \wedge \omega_3 \wedge \ldots) =$$

$$(d\omega_1 \wedge \omega_2 \wedge \omega_3 \wedge \ldots) + ((-1)^{d_1}(\omega_1 \wedge d\omega_2 \wedge \omega_3 \wedge \ldots) \qquad (12)$$

$$+ (-1)^{d_1 + d_2}(\omega_1 \wedge \omega_2 \wedge d\omega_3 \wedge \ldots) + \ldots$$

where $d_i = \deg \omega_i$. This is easily established, and we leave it to the reader.

Without doubt the most important single formula of the calculus of differential forms is

$$\boxed{d^2 = 0} \; , \qquad (13)$$

which means $d(d\omega) = 0$ for any form ω. To establish this it suffices, by (3), to consider a form ω of one component,

$$\omega = a \, dx_{i_1} \wedge \ldots \wedge dx_{i_r} . \qquad (14)$$

Taking the order of subscripts as they come, we have

$$d\omega = (\sum_j a_j dx_j) \wedge dx_{i_1} \wedge \ldots \wedge dx_{i_r} ,$$

$$d(d\omega) = \sum_j (\sum_i a_{ji} dx_i) \wedge dx_j \wedge dx_{i_1} \wedge \ldots \wedge dx_{i_r}$$

$$= \left\{ \sum_{i<j} (a_{ji} - a_{ij}) dx_i \wedge dx_j \right\} \wedge dx_{i_1} \wedge \ldots \wedge dx_{i_r} .$$

By the equality of mixed partials the form in brackets vanishes, whence so does $d(d\omega)$, QED.

This argument has another version which is instructive. Put $\tau = dx_{i_1} \wedge \ldots \wedge dx_{i_r}$. Since its coefficient function is a constant (identically 1) we have $d\tau = 0$. The form ω of (14) is $\omega = a \wedge \tau$. By Leibniz's rule,

$$d\omega = da \wedge \tau + a \wedge d\tau = da \wedge \tau ,$$

$$d^2\omega = (d^2a) \wedge \tau + da \wedge d\tau = (d^2a) \wedge \tau ;$$

and $d^2a = d(\sum_j a_j dx_j) = \sum_{i<j} (a_{ji} - a_{ij}) dx_i \wedge dx_j = 0$, hence $d^2\omega = 0$.

By Leibniz's rule and (13),

$$d(\omega \wedge d\tau) = d\omega \wedge d\tau , \tag{15}$$

for $d(\omega \wedge d\tau) = d\omega \wedge d\tau + (-1)^{\deg \omega} \omega \wedge d^2\tau$. The special case

$$d(\omega \wedge d\omega) = d\omega \wedge d\omega \tag{16}$$

is noteworthy. Together with the Corollary to 2.3(13) it yields

$$\deg \omega \quad \text{even} \implies d(\omega \wedge d\omega) = 0 . \tag{17}$$

Here is another way to get (17). By the commutation rule, $d\omega \wedge \omega = (-1)^{r(r+1)} \omega \wedge d\omega$, where $r = \deg \omega$. Now $r(r+1)$ is even for any integer r, whence for any form ω

$$d\omega \wedge \omega = \omega \wedge d\omega . \tag{18}$$

Then by Leibniz's rule, $d(\omega \wedge \omega) = \{1 + (-1)^{\deg \omega}\}\omega \wedge d\omega ,$ hence

$$\deg \omega \quad \text{even} \implies \omega \wedge d\omega = \tfrac{1}{2}d(\omega \wedge \omega) . \tag{19}$$

By (13) this implies (17). The case of (19) in which $\deg \omega = 0$, which is to say $\omega = f \in \Lambda^0$, is the familiar formula

$$d(f^2) = 2f \, df . \tag{20}$$

4.2 Fundamental Theorem of Calculus

The r-dimensional unit "cube" B_r in $\mathbb{R}^r$ is

$$B_r : 0 \le x_i \le 1 , \quad i = 1, 2, \ldots, r . \tag{1}$$

Its "faces" are of dimension $r-1$, a typical one being

$$F^1 : \begin{cases} x_1 = 1 \\ 0 \le x_i \le 1, \quad i = 2, 3, \ldots, r . \end{cases} \tag{2}$$

The face opposite to F^1 is

$$G^1 : \begin{cases} x_1 = 0 \\ 0 \le x_i \le 1, \quad i = 2, 3, \ldots, r . \end{cases} \tag{3}$$

We unify the notation by relabelling (2), (3) as F_1^1, F_0^1 respectively. There are $2r$ faces, in r opposite pairs

$$F_0^1, F_1^1 ; F_0^2, F_1^2 ; \ldots , F_0^r, F_1^r . \tag{4}$$

The collection of all faces (4) of B_r is called the <u>boundary</u> of B_r, and denoted ∂B_r. We observe that ∂B_r divides $\mathbb{R}^r$ into two regions, the inside and the outside of B_r.

We have said that an orientation is given to the volume element on a parametrically given manifold M by the ordering of the parameters. As a geometric consequence, a small change in the parameters causes a small change in the tangent space and in the volume "vector" 3.4(7). In a word, the tangent space and normal space at a point turn smoothly as the point moves on M. If M is of dimension $(r-1)$ in $\mathbb{R}^r$, the normal space is of dimension 1, so an orientation means that the normal vector has the same "sense" at every point of M: if M has an outside and an inside, the normal vector points always outward or always inward.

Along the edges of ∂B_r there is no tangent space. But if we image ∂B_r rounded off, we can then orient it by requiring the normal to be always outward, say; then, by regarding ∂B_r as the limiting case of the rounded-off ∂B_r, we find a consistent orientation for ∂B_r: the normal on each face F_0^j, F_1^j should be outward, say. This dictates the choice of the volume element on each face. Using the Cartesian co-ordinates in $\mathbb{R}^r$ as parameters for ∂B_r we may proceed as follows: on each face we take the wedge product of the differentials, with subscripts

in increasing order, with one differential omitted, and with algebraic
sign undetermined; we then let a generic point move smoothly along a
closed curve on ∂B_r, and choose the algebraic signs so that the forward
motion of the point accords with the increase or decrease of the appropriate variables on each face. By choosing this path to be an equatorial
one, so that its tangent vector is one of the basic vectors in the tangent
space according to this parametrization, we see that this method of
choosing signs forces a consistent choice of orientation on each face of
B_r. We illustrate the procedure in low dimensions, including for
completeness the degenerate case of

B_1 in $\mathbb{R}^1$.

Example 1. In $\mathbb{R}^1$ the unit "cube"
B_1 is the unit interval, and ∂B_1

is the two-point set $\{0,1\}$. ∂B_1 divides $\mathbb{R}^1$ into the inside and
outside of B_1. There is no reasonable tangent space to B_1, but we
may supply an orientation by choosing signs at the endpoints as indicated,
and we are free to think of this as giving the outward sense of the
"normal vector".

Example 2. In $\mathbb{R}^2$ the unit "cube"
B_2 is the unit square. We first
round off the corners, and then let
a point traverse the cube counter-
clockwise, choosing $\pm dx_i$, $i = 1,2$,
for the volume element on the faces
consistent with this motion. The
result is as follows:

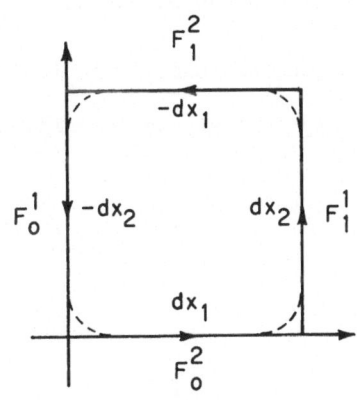

$$\text{On} \quad F_0^1 : dv = -dx_2 \ ,$$

$$\text{On} \quad F_1^1 : dv = dx_1 \ ;$$

(5)

and the volume element on the faces opposite must have opposite signs, clearly.

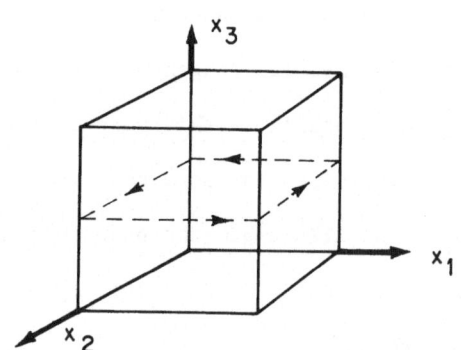

Example 3. In $\mathbb{R}^3$ the unit "cube" B_3 is the unit cube. In the figure we have indicated by a dashed line a typical equatorial path. If we let the moving point traverse it in the direction shown, then we shall have

$dx_2 \wedge dx_3$ for the volume element on F_1^1, and consequently $-dx_2 \wedge dx_3$ on the opposite face F_0^1. Proceeding in this way we find:

$$\text{On} \quad F_0^1 : dv = -dx_2 \wedge dx_3 \ ,$$

$$\text{On} \quad F_0^2 : dv = dx_1 \wedge dx_2 \ ,$$

$$\text{On} \quad F_0^3 : dv = -dx_1 \wedge dx_3 \ ;$$

(6)

with opposite signs on opposite faces.

The general rule is the literal extension of (6) to higher dimensions. Explicitly, the rule is:

$$\text{On} \quad F_0^i, \quad dv = (-1)^i dx_1 \wedge \ldots \wedge \widehat{dx_i} \wedge \ldots \wedge dx_r \ ,$$

(7)

$$i = 1, 2, \ldots, r \ ,$$

where the cap means that the term is deleted. On the opposite faces F_1^i, then,

$$dv = (-1)^{i-1} dx_1 \wedge \ldots \wedge \widehat{dx_i} \wedge \ldots \wedge dx_r \, , \tag{8}$$

$$i = 1, 2, \ldots, r \, .$$

Let ω be an $(r-1)$ form in $\mathbb{R}^r$. Since ω has r components we may use the simplified notation

$$\omega = \sum_{i=1}^{r} a^i dx_1 \wedge \ldots \wedge \widehat{dx_i} \wedge \ldots \wedge dx_r \, . \tag{9}$$

Then

$$d\omega = (\sum_{i=1}^{r} (-1)^{i-1} a_i^i) dx_1 \wedge \ldots \wedge dx_r \, . \tag{10}$$

We calculate $\int_{B_r} d\omega$. It is a sum of r-fold integrals, and in the ith term the integration on the ith coordinate can be carried out:

$$\int \ldots \int a_i^i(x_1, \ldots, x_r) dx_1 \wedge dx_2 \wedge \ldots \wedge dx_r =$$
$$\tag{11}$$
$$\int\!\!\int (a^i(x_1, \ldots, 1, \ldots, x_r) - a^i(x_1, \ldots, 0, \ldots, x_r)) dx_1 \wedge \ldots \wedge \widehat{dx_i} \wedge \ldots \wedge dx_r \, .$$

Notice the sign $(-1)^{i-1}$ in (10), and in (8); notice the minus sign in (11), and the sign $(-1)^i$ in (7). Taking these signs into account we find the remarkable fact that (11) (with sign $(-1)^{i-1}$) is the integral of ω over the two (oriented) faces F_0^i, F_1^i; whence, after the addition (10), we have the formula

$$\int_{B_r} d\omega = \int_{\partial B_r} \omega \, . \tag{12}$$

The <u>principle of orientation-cancellation</u> may be stated, for present purposes, as follows. Suppose we have two replicas of B_r, situated so that they have a face in common. Call the composite figure C_r. If τ is an (r-1) form, then

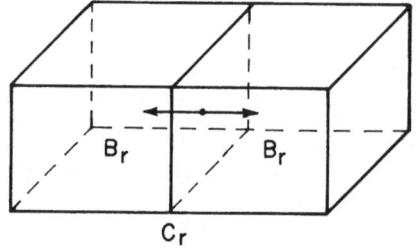

$$\int_{\partial C_r} \tau = \int_{\partial B_r} \tau + \int_{\partial B_r'} \tau \; : \quad \text{the integral}$$

over the boundary of the composite manifold is the sum of the integrals over the boundaries of the component manifolds. The reason is, that on the common face the two volume elements (one for each component) have opposite sign; and therefore the two integrals of τ over this face will cancel.

By this principle the formula (12) extends to any manifold M of dimension r in $\mathbb{R}^r$ which has a boundary and is orientable. For we may partition M into a system of small cubes, and if the mesh of this partition is sufficiently fine the system of cubes approximates M well. If we denote the cubes by $B_1, B_2, \ldots$, and if ω is an (r-1)-form, then $\int_M d\omega \underset{\sim}{} \sum_i \int_{B_i} d\omega = \sum_i \int_{\partial B_i} \omega$. By the orientation-cancellation principle, $\sum_i \int_{\partial B_i} \omega \underset{\sim}{} \int_{\partial M} \omega$. Therefore $\int_M d\omega \underset{\sim}{} \int_{\partial M} \omega$, and in the limit as the mesh of the partition is refined, we have equality,

$$\int_{\partial M} \omega = \int_M d\omega \, . \tag{13}$$

Though the result (13) is essentially correct, the foregoing

argument for it is not correct. There is a serious technical difficulty

in approximating ∂M by flat faces parallel to the coordinate "planes".[1]

The correct argument requires apparatus too far afield to be undertaken

here; we shall have to be content with the foregoing plausibility

argument.

Let M be an oriented r-manifold in $\mathbb{R}^k$, $r < k$, with boundary ∂M.

Let ω be an (r-1)-form. In the near vicinity of each of its points

M is to first approximation its tangent space $\mathbb{R}^r$. We look upon it as

being (approximately) a small r-cube at each of its points. If we restrict

the argument $\vec{x}$ to lie always in M the coefficient function

$a^{i_1, \ldots, i_r}(\vec{x})$ of an r-form ω are functions of r variables only, which

we may take locally to be the Cartesian coordinates in the local r-cube.

By (12) the integral of $d\omega$ over each local r-cube equals the integral

of ω over its boundary. By piecing together these integrals over the

local r-cubes everywhere in M we approximate, on one hand, $\int_M d\omega$, and

on the other (by orientation-cancellation) $\int_{\partial M} \omega$. Therefore the following

theorem must be correct.

__Theorem.__ If M is an oriented r-manifold in $\mathbb{R}^k$ with boundary ∂M,

and ω is an (r-1) form, $r = 1, 2, \ldots, k$, then

---

(1) This difficulty is discussed in Courant [1], volume II; see
     page 268, and also page 381 et seq. There is the further diffi-
     culty that not all manifolds have smooth boundaries; moreover,
     there exist manifolds which cannot be oriented. But all this
     must be passed over here, and we take (13) as an assertion con-
     cerning the "generic" manifold M.

$$\boxed{\int_{\partial M} \omega = \int_M d\omega} \ .$$

(14)

This is the higher dimensional generalization of the fundamental theorem of calculus: take $k = r = 1$; then M is an interval $M = (a,b)$, $\partial M = \{a,b\}$, with orientation as given in Example 1 above; for a zero form f, $\int_{\partial M} f = f(b) - f(a)$, the "integral" over the boundary is just the value of f at the two points with signs according to the orientation; and $\int_M df = \int_a^b f'(x)dx$; so (14) is in this instance

$$f(b) - f(a) = \int_a^b f'(x)dx \ .$$

Take $h = 2$, $r = 1$; that is, consider a one-form $\omega = a^1 dx_1 + a^2 dx_2$ in $\mathbb{R}^2$. If M is a planar region bounded by a closed curve ∂M, then (14) says that

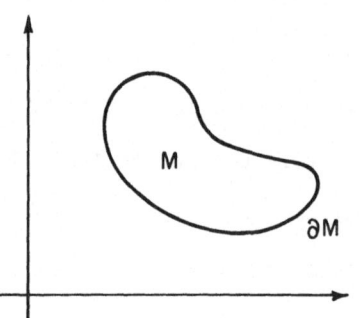

$$\int_{\partial M} a^1 dx_1 + a^2 dx_2 = \iint_M (a_1^2 - a_2^1) dx_1 dx_2 \ .$$

(15)

This formula, which is known as <u>Gauss' theorem in the plane</u>, is familiar to students of elementary differential equations.[1] Its general form (that is, for a one-form in $\mathbb{R}^k$) is

$$\int_{\partial M} \{\textstyle\sum a^i dx_i\} = \int_M \sum_{i<j} (a_i^j - a_j^i) dx_i \wedge dx_j \ ,$$

(16)

-----

(1)   See Courant [1], volume II, page 359, for the traditional proof.

where M is a surface in $\mathbf{R}^r$ bounded by a closed curve ∂M .

We shall have other such formulae when we discuss vector analysis in $\mathbf{R}^3$.

4.3 Closed Forms

A form ω is called closed if $d\omega = 0$.

Example 1. If $\omega = d\tau$ then $d\omega = d^2\tau = 0$ (4.1(13)).

Example 2. If ω and τ are closed, then $\omega \wedge \tau$ is closed (by Leibniz's rule, $d(\omega \wedge \tau) = d\omega \wedge \tau + (-1)^{\deg \omega} \omega \wedge d\tau = 0$) .

Example 3. $\omega \wedge d\omega$ is closed if ω has even degree (see 4.1(17) and 4.1(19)).

Example 4. If ω is closed, then $\omega \wedge d\tau$ is closed, for any τ (4.1(15)).

The basic property of closed forms is as follows.

Theorem. An r-form ω in $\mathbf{R}^k$ is closed if and only if

$$\int_{\partial M} \omega = 0 \tag{1}$$

for every (r+1)-manifold M in $\mathbf{R}^k$.

Proof: Assume (1). We may take for M any (r+1)-parallelopiped, as small as we like, and involving any (r+1) of the k coordinates, containing a given point $\vec{a}$ of $\mathbf{R}^k$. By the Fundamental theorem 4.2(14), $\int_M d\omega = 0$ for all such M, whence $d\omega = 0$ at $\vec{a}$. Since $\vec{a}$ was chosen arbitrarily, $d\omega = 0$.

Conversely, if ω is closed, then $\int_{\partial M} \omega = \int_M d\omega = \int_M 0 = 0,$ which is (1). This proves the theorem QED.

The closure of a form may be expressed by a differential equation in its component functions:

$$\sum a^{i_1,\ldots,i_r} dx_{i_1} \wedge \ldots \wedge dx_{i_r} \quad \text{is closed} \iff$$

$$\sum_{\nu=0}^{r} (-1)^{\nu} a_{i_\nu}^{i_0,\ldots,\hat{i}_\nu,\ldots,i_r} = 0, \quad \text{all} \quad i_0 < \ldots < i_r . \tag{2}$$

This follows at once from the general formula 4.1(4) for the exterior derivative of an r-form. Here are a few special cases of (2).

<u>Example</u> 5. A zero-form f is closed if and only if f is constant (here (2) says $f_i = 0$, $i = 1, 2, \ldots, k$).

<u>Example</u> 6. A one-form $\omega = \sum a^i dx_i$ is closed if and only if

$$a_i^j = a_j^i, \quad \text{all} \quad i < j . \tag{3}$$

<u>Example</u> 7. A two-form $\sum_{i<j} a^{i,j} dx_i \wedge dx_j$ in $\mathbb{R}^3$ is closed if and only if

$$a_1^{23} - a_2^{13} - a_3^{12} = 0 . \tag{4}$$

Motivated by (3), we call the equation (2), second line, the condition of <u>index symmetry</u>.

Let a path in $\mathbb{R}^k$ be given parametrically as

$$\vec{\gamma}(t), \quad a \le t \le b . \tag{5}$$

Define a new function $\tilde{\vec{\gamma}} : \mathbb{R}^1 \to \mathbb{R}^k$ by

$$\sim\vec{\gamma}(t) = \vec{\gamma}(t-b+a), \quad a \le t \le b . \tag{6}$$

$\sim\vec{\gamma}$ is $\vec{\gamma}$ with the opposite orientation (or, traversed backwards). For any one-form ω we have

$$\int_{\sim\vec{\gamma}} \omega = -\int_{\vec{\gamma}} \omega . \tag{7}$$

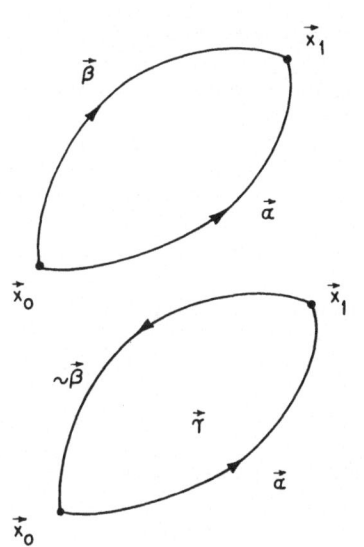

Let $\vec{\alpha}, \vec{\beta}$ be two curves in $\mathbb{R}^k$ going from a given initial point $\vec{x}_0$ to a given endpoint $\vec{x}_1$ (see Figure). Then $\vec{\alpha}, \sim\vec{\beta}$ together constitute a closed curve $\vec{\gamma}$. Let M be a 2-manifold formed by filling in the "inside" of $\vec{\gamma}$ in any smooth way: one may imagine gluing a thin rubber sheet to $\vec{\gamma}$, for instance. Then $\vec{\gamma} = \partial M$.

If ω is a closed one-form, then $\int_{\partial M} \omega = \int_{\vec{\gamma}} \omega = 0$, which is to say, $\int_{\vec{\alpha}} \omega + \int_{\sim\vec{\beta}} \omega = 0$, whence by (7)

$$\int_{\vec{\alpha}} \omega = \int_{\vec{\beta}} \omega . \tag{8}$$

We have shown that the value of the path integral of a closed one-form depends upon the initial and endpoints only, and is independent of any twistings and turnings which may occur between. Call this the property of <u>path-independence</u> for ω . By reversing the foregoing argument we find that if ω has path-independence then its integral over any closed curve vanishes, whence (by the theorem) ω is closed. We state this result as the following

Corollary. A one-form is closed if and only if it has the property

                 of path-independence.

     Here is an elementary proof of the second half of the Corollary (path-independence implies closure). Let P denote $\vec{x}$, let P_i denote $\vec{x}$ with x_i replaced by 0, and so on. Given a path-independent one-form $\omega = \sum a^i dx_i$, we shall get the same value if we integrate ω along the two paths (P_{ij}, P_j, P) and (P_{ij}, P_i, P).

Since there are only the two operative variables x_i, x_j, we suppress the others and write the equality thus:

$$\int_0^{x_i} a^i(t,0)\,dt + \int_0^{x_j} a^j(x_i,s)\,ds = \int_0^{x_j} a^j(0,u)\,du + \int_0^{x_i} a^i(v,x_j)\,dv . \tag{9}$$

Differentiation through (9) by x_j yields

$$a^j(x_i,x_j) = a^j(0,x_j) + \int_0^{x_i} a^i_j(v,x_j)\,dv ; \tag{10}$$

differentiation through (10) by x_i yields

$$a^j_i(x_i,x_j) = a^i_j(x_i,x_j) . \tag{11}$$

This is index symmetry (3) for ω, which therefore implies its closure, QED.

4.4   Exact Forms

A form ω is called __exact__ if there is another form τ (of one degree less) such that $\omega = d\tau$.

__Example 1.__ $\sum_i f_i dx_i$ is exact, being equal to df.

__Example 2.__ $\sum_{i<j} (a_i^j - a_j^i) dx_i \wedge dx_j$ is exact, being equal to $d(\sum a^i dx_i)$.

__Example 3.__ If ω has even degree, then $\omega \wedge d\omega$ is exact, $\omega \wedge d\omega = d(\frac{1}{2}\omega \wedge \omega)$ (4.1(19)).

__Example 4.__ $d\omega \wedge d\tau$ is exact for any ω, τ; $d\omega \wedge d\tau = d(\omega \wedge d\tau)$ (4.1(15)).

An exact form is closed: if $\omega = d\tau$ then $d\omega = d^2\tau = 0$ (4.1(13)). One can see this in Example 1 without relying upon 4.1(13) by observing that $d(\sum_i f_i dx_i)$ has index symmetry; and similarly in Example 2, by a longer calculation. Such calculations amount to pieces of a clumsy proof that $d^2 = 0$.

The reader should recall at this time the following matter about first order differential equations. The equation

$$M(x,y)dx + N(x,y)dy = 0 \qquad (1)$$

is said to be an "exact equation" if there is a function $\phi(x,y)$ such that $\frac{\partial}{\partial x}\phi = M$, $\frac{\partial}{\partial y}\phi = N$, so that (1) is equivalent to the integrable equation $d\phi = 0$. From our present vantage point the equation (1) is "exact" if and only if its left side is an exact one-form. In the elementary texts one goes on to prove that (1) is an "exact equation" if and only if $\frac{\partial}{\partial x}N = \frac{\partial}{\partial y}M$. From our present vantage point this is the statement that the left side of (1) is an exact one-form if and only if it has index symmetry; that is to say, if and only if it is closed. Part of this (that exact forms are automatically closed) we have just remarked. The other part (that if the left side of (1) is a closed one-form then (1) is "exact") is an instance of the remarkable theorem that every closed form is exact. To put it more forcefully: not only are exact forms closed, but the <u>only</u> way a form can be closed is to be exact. This can be demonstrated for one-forms quite simply, as follows.

Let $\omega = \sum a^i dx_i$ be a closed one-form in $\mathbf{R}^k$. Let γ be the piece-wise rectilinear path from $\vec{0}$ to $\vec{x}$ which traverses successive edges of the k-parallelopiped with opposite corners at $\vec{0}$ and $\vec{x}$ respectively (see the Figure for the $\mathbf{R}^3$ case). Since ω is path-independent, $\int_\gamma \omega$ defines a function (zero-form) f by the equation $f(\vec{x}) = \int_\gamma \omega$. In coordinates,

$$f(\vec{x}) \ = \ \sum_{i=1}^{k} \int_{a}^{x_i} a^i(x_1, x_2, \ldots, x_{i-1}, t_i, 0, \ldots, 0)\, dt_i \ . \qquad (2)$$

Differentiating through (2) by x_k we have at once. $f_k(\vec{x}) = a^k(\vec{x})$. Differentiation of (2) by x_{k-1} yields

$$f_{k-1}(\vec{x}) \ = \ a^{k-1}(x_1, \ldots, x_{k-1}, 0) + \int_{0}^{x_k} a^k_{k-1}(x_1, \ldots, x_{k-1}, t_k)\, dt_k \ . \qquad (3)$$

By the index symmetry 4.3(3) of ω we can interchange k and $(k-1)$ in the last term; the integration there can then be carried out; the result is

$$\int_0^{x_k} a^k_{k-1}(x_1,\ldots,x_{k-1},t_k)\,dt_k \;=\; a^{k-1}(\vec{x}) - a^{k-1}(x_1,\ldots,x_{k-1},0). \qquad (4)$$

Putting (4) into (3) we find $\quad f_{k-1}(\vec{x}) = a^{k-1}(\vec{x}).\quad$ Continuing in this way we have, after k steps,

$$f_j(\vec{x}) \;=\; a^j(\vec{x}), \qquad j = 1,2,\ldots,k, \qquad\qquad (5)$$

or $\quad \omega = df \quad$ is exact, QED.

We have here solved the <u>form differential equation</u> $\quad df = \omega \quad$ for the unknown $\quad f \in \Lambda^0 \quad$ given $\quad \omega \in \Lambda^1 \quad$ closed. The solution (2) is unique up to an additive closed zero-form, for if g is a second solution, then $d(f-g) = \omega - \omega = 0$. In components the equation $\quad df = \omega \quad$ is the first order system $\quad f_i = a^i,\quad i = 1,2,\ldots,k,\quad$ of partial differential equations, where $\quad \omega = \sum a^i dx_i,\quad$ and we used the index symmetry $\quad a^i_j = a^j_i \quad$ in solving the system. The higher order form-differential equation $\quad d\tau = \omega \quad$ for an unknown $\quad \tau \in \Lambda^{r-1} \quad$ given $\quad \omega \in \Lambda^r \quad$ is similarly a first order system of partial differential equations, though rather more complicated than the foregoing case. Its solutions are unique up to an additive closed $(r-1)$-form, for if σ and τ are solutions, so that $\quad d\sigma = \omega = d\tau,$ then $\quad d(\sigma-\tau) = 0.\quad$ If ω is closed, then one can show, by means of the index symmetry 4.3(2), that the system has solutions. This method of solution is in general laborious, though for low degrees it is feasible. The case of a closed two-form in $\mathbb{R}^3$ is treated this way in texts on vector analysis, where it is expressed as the theorem that a source-free vector field is a curl.[1]   Together with our treatment above of closed one-forms, this supplies what we will need in the following chapters: namely, the facts that closed one-forms and two-forms in $\mathbb{R}^3$ are exact; and we forego the discussion here of the general theorem.

_____

(1)   This terminology will be explicated in Chapter 5, or one may consult Courant [1], volume II. See also page 404 there for a proof of the type indicated for the theorem under discussion.

Chapter 5
Vector operations in $\mathbb{R}^3$

5.1 Nabla

In the vector analysis of $\mathbb{R}^3$ one uses a formal vector operator, written ∇ and called "nabla" or "del", which is defined (so to speak) as

$$\nabla = \sum_{i=1}^{3} \frac{\partial}{\partial x_i} \vec{e}_i . \tag{1}$$

Its action on a scalar field is

$$\nabla f = \sum f_i \vec{e}_i ; \tag{2}$$

that is, $\nabla f = \vec{g}$, where $g^i = f_i$. Note that ∇f is a vector field. It is also denoted grad f, read "gradient of f". There are two ways for ∇ to act on a vector field $\vec{f}$:

$$\nabla \cdot \vec{f} = \sum f_i^i , \tag{3}$$

$$\nabla \times \vec{f} = (f_2^3 - f_3^2) \vec{e}_1 + (f_3^1 - f_1^3) \vec{e}_2 + (f_1^2 - f_2^1) \vec{e}_3 . \tag{4}$$

$\nabla \cdot \vec{f}$ is a scalar field, also denoted $\text{Div } \vec{f}$, read "divergence of $\vec{f}$".

$\nabla \times \vec{f}$ is a vector field, also denoted $\text{Curl } \vec{f}$, read "curl of $\vec{f}$". The

dot and cross of (3) and (4) are "formal" scalar and vector products in

the following senses: $\nabla \cdot \vec{f} = \dfrac{\partial}{\partial x_1} f^1 + \dfrac{\partial}{\partial x_2} f^2 + \dfrac{\partial}{\partial x_3} f^3$; and $\nabla \times \vec{f}$ is

the result of the formal evaluation of the determinant

$$\begin{vmatrix} \vec{e}_1 & \vec{e}_2 & \vec{e}_3 \\ \dfrac{\partial}{\partial x_1} & \dfrac{\partial}{\partial x_2} & \dfrac{\partial}{\partial x_3} \\ f^1 & f^2 & f^3 \end{vmatrix} \ .$$

With this in mind, the notations (3), (4) for Div, Curl make it auto-

matically clear that $\text{Div } \vec{f}$ is a scalar field and $\text{Curl } \vec{f}$ is a

vector field.

This notation will appeal to anyone who has a taste for puns, and

there is really nothing wrong with it. However, it is possible, indeed

it is easy, to identify nabla and its manifestations as recognizable

entities in the apparatus of differential forms, and thereby greatly to

simplify many of the arguments and calculations of vector analysis. To

this we now turn.

In $\mathbb{R}^3$ we have the form spaces $\Lambda^0, \Lambda^1, \Lambda^2, \Lambda^3$ of dimensions

$1, 3, 3, 1$ respectively. In particular, the dimensions of Λ^1 and Λ^2

are both equal to that of the underlying space $\mathbb{R}^3$. This is a fact

peculiar to $\mathbb{R}^3$: In $\mathbb{R}^k$, the dimension of Λ^1 is $\binom{k}{1} = k$, but the

dimension $\binom{k}{2}$ of Λ^2 is k only when $k = 3$. (1) This peculiar fact

_____

(1) $\binom{k}{2} = \dfrac{k(k-1)}{2} = k$ implies $k-1 = 2$, $k = 3$.

makes it possible to identify (set up a one-to-one correspondence between) $\wedge^1$ and $\wedge^2$. The space of all vector fields, which we shall denote by W, is also a vector space of dimension 3 with scalar-field coefficients: its basis is $\vec{e}_1, \vec{e}_2, \vec{e}_3$; its typical element is $\vec{f} = \sum f^i \vec{e}_i$, where f^i are scalar fields. Therefore we are able, in $\mathbb{R}^3$ only, to interpret a vector field either as a one-form or as a two-form. Since the wedge product of one-forms is a two-form, and the vector product of vector fields is a vector field, we want to set up the correspondences between $\wedge^1$, $\wedge^2$ and W in such a way that the wedge and vector products correspond. One sees how to do this at once by comparing the two product formulae:

$$\vec{f} \times \vec{g} = (f^2 g^3 - f^3 g^2)\vec{e}_1 + (f^3 g^1 - f^1 g^3)\vec{e}_2 + (f^1 g^2 - f^2 g^1)\vec{e}_3 ,$$

$$(\sum a_i dx_i) \wedge (\sum b^i dx_i) = (a^2 b^3 - a^3 b^2) dx_2 \wedge dx_3$$
$$+ (a^1 b^3 - a^3 b^1) dx_1 \wedge dx_3 + (a^1 b^2 - a^2 b^1) dx_1 \wedge dx_2 .$$

The correspondences are given in the following table.

W	$\wedge^1$	$\wedge^2$
$\vec{e}_1$	dx_1	$dx_2 \wedge dx_3$
$\vec{e}_2$	dx_2	$-dx_1 \wedge dx_3$
$\vec{e}_3$	dx_3	$dx_1 \wedge dx_2$

(5)

The minus sign in the second line derives from the fact that the subscripts in a two-form come in increasing (rather than cyclic) order.

The set of all scalar fields, which we denote by $, is a vector
space of dimension 1 with scalar-field coefficients: for its basis we
take the scalar 1, and its typical element is $f = f \cdot 1$.   We may therefore
identify $ with $\wedge^0$ and $\wedge^3$:

$	$\wedge^0$	$\wedge^3$
1	1	$dx_1 \wedge dx_2 \wedge dx_3$

(6)

By means of the correspondences (5), (6) we may interprete the nabla
operations as follows.

<u>Grad</u>: A scalar field f is via (6) a zero-form; the one-form $df = \sum f_i dx_i$

is via (5) the vector field $\nabla f = \sum f_i \vec{e}_i$.   This is (2).

<u>Curl</u>: A vector field $\vec{f}$ is via (5) a <u>one-form</u> $\vec{f} = \sum f^i dx_i$;   the

two-form $d\vec{f} = \sum_{i<j} (f_i^j - f_j^i) dx_i \wedge dx_j$ is via (5) the vector field

$\nabla \times \vec{f}$.   This is (4).

<u>Div</u>:   A vector field $\vec{f}$ is via (5) a <u>two-form</u> $\vec{f} = f^1 dx_2 \wedge dx_3 - f^2 dx_1 \wedge dx_3 +$

$f^3 dx_1 \wedge dx_2$;   the three-form $d\vec{f} = (\sum f_i^i) dx_1 \wedge dx_2 \wedge dx_3$ is via (6)

the scalar field $\nabla \cdot \vec{f}$.   This is (3).

Thus all three operations are exterior differentiation. Note the
dichotomy   Curl $\vec{f}$, Div $\vec{f}$ according as $\vec{f}$ is viewed as a one-form or
as a two-form. We summarize:

$$\text{Grad } f = df \qquad f \text{ in } \wedge^0 ,$$
$$\text{Curl } \vec{f} = d(\vec{f}), \quad \vec{f} \text{ in } \wedge^1 ,$$
$$\text{Div } \vec{f} = d(\vec{f}), \quad \vec{f} \text{ in } \wedge^2 ,$$

(7)

where we write "f in Λ^1" for "f interpreted by (5), (6) as a member of Λ^1". In the same spirit

$$\vec{f} \times \vec{g} = \vec{f} \wedge \vec{g}, \quad \vec{f} \text{ and } \vec{g} \text{ in } \Lambda^1,$$

$$\vec{f} \cdot \vec{g} = \vec{f} \wedge \vec{g}, \quad \vec{f} \text{ in } \Lambda^1, \ \vec{g} \text{ in } \Lambda^2. \tag{8}$$

Thus the exterior derivative encompasses the three nabla operations, and the wedge product expresses the scalar and vector products.

Many of the familiar formulae of vector analysis are nothing but Leibniz's rule for forms. We shall treat the following products:

$$fg, \quad \vec{f} \cdot \vec{g}, \quad f\vec{g}, \quad \vec{f} \times \vec{g}. \tag{1} \tag{9}$$

For zero-forms f and g, $d(fg) = df \wedge g + f \wedge dg$, so by (7)

$$\text{Grad}(fg) = (\text{Grad } f)g + f(\text{Grad } g). \tag{10}$$

Since $\vec{f} \cdot \vec{g}$ is a sum of such products, $\text{Grad}(\vec{f} \cdot \vec{g})$ is a sum of terms like (10). Given f and $\vec{g}$, $d(f\vec{g}) = (df) \wedge \vec{g} + f \wedge d\vec{g}$ for both interpretations of $\vec{g}$. If $\vec{g}$ is in Λ^1 the formula becomes (by (7), (8))

$$\text{Curl}(f\vec{g}) = (\text{Grad } f) \times \vec{g} + f(\text{Curl } \vec{g}) \tag{11}$$

and if $\vec{g}$ is in Λ^2 then it is

---

(1) The only other possibilities are $f \circ g$, $\vec{f} \circ \vec{g}$. Both are rejected because they do not distribute from the left across sums: $f \circ (g+h)$ will in general differ from $(f \circ g) + (f \circ h)$ when f is not linear, as it typically is not; and the same for $f \circ (\vec{g} + h)$.

$$\text{Div}(f\vec{g}) \;=\; (\text{Grad } f)\cdot\vec{g} + f(\text{Div }\vec{g}) . \qquad (12)$$

Given $\vec{f}$ and $\vec{g}$, there are two cases: $\vec{f}$ and $\vec{g}$ both in Λ^1; or one in Λ^1, the other in Λ^2. In the second case, $\vec{f}\wedge\vec{g}$ is in Λ^3, so $d(\vec{f}\wedge\vec{g}) = 0$, $d\vec{f}\wedge\vec{g} = 0$, $\vec{f}\wedge d\vec{g} = 0$. In the first case, Leibniz's rule gives $d(\vec{f}\wedge\vec{g}) = d\vec{f}\wedge\vec{g} - \vec{f}\wedge d\vec{g}$, or

$$\text{Div}(\vec{f}\times\vec{g}) \;=\; \text{Curl }\vec{f}\cdot\vec{g} - \vec{f}\cdot\text{Curl }\vec{g} . \qquad (13)$$

There is one more case. Given $\vec{f}$ and $\vec{g}$ interpreted as one-forms, we may interprete the two-form $\vec{f}\wedge\vec{g}$ as the one-form

$$\tau \;=\; (f^2 g^3 - f^3 g^2)\,dx_1 - (f^1 g^3 - f^3 g^1)\,dx_2 - (f^1 g^2 - f^2 g^1)\,dx_3 , \qquad (14)$$

and then take $d\tau$. This will yield $\text{Curl}(\vec{f}\times\vec{g})$. Abbreviate τ as $\tau = \sum A^i dx_i$. The first component of $d\tau$ will then be $A^3_2 - A^2_3$. By (14) this is

$$
\begin{aligned}
A^3_2 - A^2_3 \;=\;& f^1_2 g^2 + f^1 g^2_2 - f^2_2 g^1 - f^2 g^1_2 \\[4pt]
& + f^1_3 g^3 + f^1 g^3_3 - f^3_3 g^1 - f^3 g^1_3 .
\end{aligned}
\qquad (15)
$$

The other components of $\text{Curl}(\vec{f}\times\vec{g})$ may be deduced from (15) by permuting the indices.

Let $\vec{r}$ be the position vector field,

$$r^i(\vec{x}) \;=\; x_i, \qquad i = 1,2,3 . \qquad (16)$$

If $\vec{\omega}$ is a constant vector field then

$$\text{Curl}(\vec{\omega} \times \vec{r}) = 2\vec{\omega} ; \tag{17}$$

for, putting $\vec{\omega}$ for $\vec{f}$ and $\vec{r}$ for $\vec{g}$ in (15), all terms involving a derivative of f vanish, and $g_j^i = \delta_{ij}$, so what remains is $2f^1 = 2\omega^1$; and similarly for the other components. We will need (17) later on.

The directional derivative $D_{\vec{u}}f$ of a scalar field in the direction of $\vec{u}$ is also connected with nabla. The definition is

$$D_{\vec{u}}f(\vec{x}) = \lim_{t \to 0} \frac{f(\vec{x} + t\vec{u}) - f(\vec{x})}{t} \tag{18}$$

which may be rephrased as

$$D_{\vec{u}}f(\vec{x}) = \frac{d}{dt} f(\vec{x} + t\vec{u}) \bigg|_{t=0} . \tag{19}$$

By the chain rule this is $\sum_i f_i(\vec{x} + t\vec{u}) u_i \bigg|_{t=0} = \sum f_i(\vec{x}) u_i$, whence

$$D_{\vec{u}}f = \text{Grad } f \cdot \vec{u}. \tag{20}$$

Note the special case

$$D_{\vec{e}_i} f = f_i . \tag{21}$$

Take $\vec{u}$ to be a unit vector. Then $|D_{\vec{u}}f| \le \|\text{Grad } f\|$ by (20). Since

$$\sum_i f_i \cdot \frac{f_i}{\sqrt{\sum_j f_j^2}} = \sqrt{\sum_i f_i^2} = \|\text{Grad } f\|,$$

we conclude that the maximum value of $|D_{\vec{u}} f|$ is $\|\text{Grad } f\|$, and that this maximum is achieved when $\vec{u}$ is parallel to Grad f.

5.2 Higher Derivatives

Since Grad, Div, and Curl are differentiations, we regard iterations of these operations as higher derivatives. The first order derivatives act as follows:

$$\text{Grad} : \mathbb{S} \to \mathbb{W},$$

$$\text{Curl} : \mathbb{W} \to \mathbb{W}, \tag{1}$$

$$\text{Div} : \mathbb{W} \to \mathbb{S}.$$

Therefore the second order possibilities are:

$$\text{Curl Grad} : \mathbb{S} \to \mathbb{W},$$

$$\text{Div Curl} : \mathbb{W} \to \mathbb{S},$$

$$\text{Curl Curl} : \mathbb{W} \to \mathbb{W}, \tag{2}$$

$$\text{Grad Div} : \mathbb{W} \to \mathbb{W},$$

$$\text{Div Grad} : \mathbb{S} \to \mathbb{S}.$$

Given f, Grad f = df is a one-form, whence Curl Grad f = $d^2 f = 0$, or

$$\text{Curl Grad} = 0 \, . \qquad (3)$$

Given $\vec{f}$ interpreted as a one-form, Curl $\vec{f} = d(\vec{f})$ is a two-form, whence Div Curl $\vec{f} = d^2(\vec{f}) = 0$, or

$$\text{Div Curl} = 0 \, . \qquad (4)$$

Statements (3) and (4) are important theorems in vector analysis. This disposes of the first two items in (2). Passing over the next two for the moment,[1] we turn to the last. Given f, Grad f must be interpreted as the two-form

$$\omega = \text{Grad } f = f_1 dx_2 \wedge dx_3 - f_2 dx_1 \wedge dx_2 + f_3 dx_1 \wedge dx_2$$

in order to apply Div.[2] Then Div Grad f = dω,

$$\text{Div Grad } f = \sum f_{ii} \, . \qquad (5)$$

This is known as the Laplace operator, or briefly the Laplacian, and is denoted Δ :

$$\Delta f = \text{Div Grad } f = \sum f_{ii} \, . \qquad (6)$$

The Laplacian has a wide-spread and far-reaching importance: in

---

(1) They will receive mention in the Appendix.

(2) This is why Div Grad is not a case of d^2.

particular it is fundamental in mathematical physics, in probability theory, and in the theory of functions of a complex variable. We must pass over all of that here.

The effect of (3), (4) is, that there are no non-zero composites of Curl with the two differentiations. Therefore all non-zero higher derivatives must be iterates of Curl alone on one hand, or composites of the other two on the other. According to (1), the latter are to be read off the following diagram:

$$\ldots \xrightarrow{\text{Grad}} \mathbb{W} \xrightarrow{\text{Div}} \mathbb{S} \xrightarrow{\text{Grad}} \mathbb{W} \xrightarrow{\text{Div}} \mathbb{S} \xrightarrow{\text{Grad}} \ldots$$

In forming such composites we write Δ for Div Grad always. A catalog of all possible non-zero derivatives is easily assembled, after these remarks, and is presented in the following table.

Type	Formula	Order	
$\mathbb{S} \to \mathbb{S}$	Δ^n, $\quad n = 1, 2, \ldots$	$2n$	
$\mathbb{W} \to \mathbb{W}$	Curl^n, $\quad n = 1, 2, \ldots$	n	
$\mathbb{W} \to \mathbb{W}$	$\text{Grad } \Delta^n \text{Div}$, $\quad n = 0, 1, \ldots$	$2n+2$	(7)
$\mathbb{S} \to \mathbb{W}$	$\text{Grad } \Delta^n$, $\quad n = 0, 1, \ldots$	$2n+1$	
$\mathbb{W} \to \mathbb{S}$	$\Delta^n \text{Div}$, $\quad n = 0, 1, \ldots$	$2n+1$	

For the second-order operator Δ we have the product rule

$$\Delta(fg) \;=\; (\Delta f)g + 2(\text{Grad } f) \cdot (\text{Grad } g) + f(\Delta g), \tag{8}$$

reminiscent of the formula

$$(fg)'' \;=\; (f'')g + 2f'g' + f(g'')$$

of one-variable calculus.

By 5.1(7)

$$\vec{f} \text{ closed} \Longleftrightarrow \left\{ \begin{array}{ll} \text{Curl } \vec{f} = 0, & f \in \Lambda^1 \\[2ex] \text{Div } \vec{f} = 0, & f \in \Lambda^2 \end{array} \right\} . \tag{9}$$

By the exactness theorem (see 4.4), in the first case $\vec{f} = dg$, $g \in \Lambda^0$, which is to say $\vec{f} = \text{Grad } g$ (see 5.1(7)); and in the second case $\vec{f} = d\vec{g}$, $\vec{g} \in \Lambda^1$, which is to say $\vec{f} = \text{Curl } \vec{g}$. These statements are companions to (3), (4) respectively. That is: not only is Curl Grad $g = 0$ for all $g \in \mathbf{S}$, but the only way Curl $\vec{f}$ can vanish is if $\vec{f} = \text{Grad } g$ for some $g \in \mathbf{S}$; and not only is Div Curl $\vec{g} = 0$ for all $\vec{g}$, but the only way Div $\vec{f}$ can vanish is if $\vec{f} = \text{Curl } \vec{g}$ for some $\vec{g}$. In the language of vector analysis these facts may be stated as the following

Theorem 1.   A curl-free field is a gradient, and a divergence-free

field is a curl.

If $\vec{f}$ and $\vec{g}$ are gradients, they are closed one-forms, whence $\vec{f} \wedge \vec{g}$ is a closed two-form (4.3, Example 2). By 5.1(8), $\vec{f} \wedge \vec{g} = \vec{f} \times \vec{g}$; whence, by 4.4, we have

Theorem 2. The vector product of gradients is a Curl.

Example 1. Let $\vec{f} = \vec{r}$, the position vector field 5.1(17). As a one-form,
$\vec{f} = x_1 dx_1 + x_2 dx_2 + x_3 dx_3$, whence $d(\vec{f}) = 0$,

$$\text{Curl } \vec{r} = 0 . \tag{10}$$

Now $\|\vec{r}\| = \sqrt{x_1^2 + x_2^2 + x_3^2} = r$, and

$$\vec{r} = \tfrac{1}{2} \text{Grad } r^2 . \tag{11}$$

As a two-form, $\vec{f} = x_1 dx_2 \wedge dx_3 - x_2 dx_1 \wedge dx_3 + x_3 dx_1 \wedge dx_2$, whence $d(\vec{f}) =$
$3 dx_1 \wedge dx_2 \wedge dx_3$,

$$\text{Div } \vec{r} = 3 . \tag{12}$$

Example 2. Let $\vec{f}_{\hat{u}}(\vec{x}) \equiv \hat{u}$, $\hat{u}$ a fixed unit vector. Looked upon as a flow, this field is a parallel flow of uniform velocity 1. As a one-form
$\vec{f}_{\hat{u}} = u^1 dx_1 + u^2 dx_2 + u^3 dx_3$, whence $d(\vec{f}_{\hat{u}}) = 0$, or

$$\text{Curl } \vec{f}_{\hat{u}} = 0 . \tag{13}$$

Since $\vec{r} \cdot \hat{u} = x_1 u^1 + x_2 u^2 + x_3 u^3$,

$$\vec{f}_{\hat{u}} = \text{Grad } \vec{r} \cdot \hat{u} . \tag{14}$$

As a two-form $\vec{f}_{\hat{u}} = u^1 dx_2 \wedge dx_3 - u^2 dx_1 \wedge dx_3 + u^3 dx_1 \wedge dx_2$, $\quad d(\vec{f}_{\hat{u}}) = 0$, or

$$\text{Div } \vec{f}_{\hat{u}} = 0 . \tag{15}$$

It will clearly be sufficient to exhibit $\vec{f}_{\hat{u}}$ as a Curl in the special case $\hat{u} = \vec{e}_1$, or $u^1 = 1$, $u^2 = u^3 = 0$. Then as a two-form $\vec{f}_{\hat{u}} = dx_2 \wedge dx_3$, and we can at once guess (by the wedge product rules) solutions to the form-differential equation $d\omega = \vec{f}_{\hat{u}}$, ω a one-form; namely $\omega_1 = x_2 dx_3$, $\omega_2 = -x_3 dx_2$ are both solutions. Since the solution is unique modulo a closed one-form, $\omega_1 - \omega_2$ has to be closed; and indeed, $\omega_1 - \omega_2 = x_3 dx_2 + x_2 dx_3$, so $d(\omega_1 - \omega_2) = -dx_2 \wedge dx_3 + dx_2 \wedge dx_3 = 0$. The average $\vec{g}$ of ω_1 and ω_2,

$$\vec{g} = -\frac{1}{2} x_3 dx_2 + \frac{1}{2} x_2 dx_3 \tag{16}$$

will also be a solution (and note that $\vec{g} - \omega_1 = -\frac{1}{2} x_3 dx_2 - \frac{1}{2} x_2 dx_3$ is closed). As a vector field $\vec{g}$ is

$$\vec{g}(x) = -\frac{1}{2} x_3 \vec{e}_2 + \frac{1}{2} x_2 \vec{e}_3 ,$$

which circulates around the x_1-axis (see Figure).

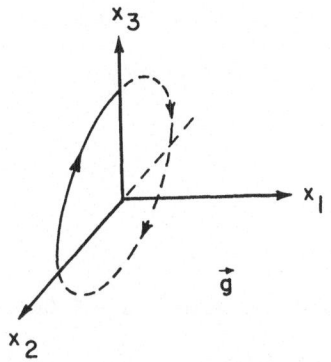

We can of course find $\vec{g}$ analytically: Let $\vec{g} = \sum a^i dx_i$, whence $d(\vec{g}) = \sum_{i<j} (a_i^j - a_j^i) dx_i \wedge dx_j$, so that $d(\vec{g}) = \vec{f}_{\hat{u}}$ is the system

$$a_2^3 - a_3^2 = 1$$

$$a_1^2 - a_2^1 = a_1^3 - a_3^1 = 0 ,$$

(17)

and a particular solution comes to hand immediately if we put $a^1 = a^2 = a_1^3 = 0$. This solution is in fact ω_1 .

Example 3. Since $\hat{u}$ and $\vec{r}$ are gradients, $\hat{u} \times \vec{r}$ must be a Curl, $\hat{u} \times \vec{r} = \text{Curl } \vec{h}$ for some $\vec{h}$. To find $\vec{h}$ we could solve the appropriate differential equation, as above. However, in this case we have a solution by the Leibniz rule formula 5.1(12): $\text{Curl}(f\vec{g}) = (\text{Grad } f) \times \vec{g} + f(\text{Curl } \vec{g})$. Put $\vec{f} = \hat{u} \cdot \vec{r}$, $\vec{g} = \vec{r}$. Then $\text{Curl}\{(\hat{u} \cdot \vec{r})\vec{r}\} = \hat{u} \times \vec{r} + (\hat{u} \cdot \vec{r})\text{Curl } \vec{r}$; since $\text{Curl } \vec{r} = 0$ (by (10)) we have

$$\hat{u} \times \vec{r} = \text{Curl}\{(\hat{u} \cdot \vec{r})\vec{r}\} .$$

(18)

Postscript: If we interprete $\hat{u}$ as the angular velocity of a rotating body, then the linear velocity $\vec{v}$ of a point $\vec{r}$ in the body is given by

$$\vec{v} = \hat{u} \times \vec{r} .$$

(19)

By 5.1(18) we have $\hat{u} = \frac{1}{2} \text{Curl}(\hat{u} \times \vec{r})$. We have therefore the following relations between $\vec{v}$ and $\hat{u}$:

$$\vec{v} = \text{Curl}\{(\hat{u} \cdot \vec{r})\vec{r}\} ,$$

$$\hat{u} = \text{Curl}\{\frac{1}{2}\vec{v}\} ;$$

(20)

and this yields an example of a Curl^2 :

$$\hat{u} = \text{Curl Curl}\{\tfrac{1}{2}(\hat{u} \cdot \vec{r})\vec{r}\}. \tag{21}$$

5.3 Integral Formulae

In 4.2 we asserted the general formula

$$\int_{M} d\omega = \int_{\partial M} \omega \tag{1}$$

and adduced Gauss' theorem in the plane 4.2(15) as a special case. In the vector analysis of $\mathbb{R}^3$ there are several other noteworthy interpretations of (1). To these we now turn.

Let f be a scalar field in $\mathbb{R}^3$, and M a curve in $\mathbb{R}^3$, $M = \{\vec{\gamma}(t); a \le t \le b\}$. Then   df = Grad f, and (1) becomes

$$\int_{a}^{b} \text{Grad } f \cdot \vec{T} \, dt = f(\vec{\gamma}(b)) - f(\vec{\gamma}(a)), \tag{2}$$

where $\vec{T}$ is the tangent to M.   This formula has the following physical interpretation. Look upon   Grad f   as a field of force. Then the line integral (2), left side, is the work integral giving the amount of work done by the field upon a test particle in moving it along the curve. Let $\phi(\vec{x}) = -f(\vec{x})$. ϕ is called the potential energy E_p of the field Grad f,   and (2) asserts that the work done equals the decrease $\phi(\vec{\gamma}(a)) - \phi(\vec{\gamma}(b))$ in the potential energy. A field of force which is the gradient of a potential function (or, which is an exact one-form) is called <u>conservative</u> for the following reason. The kinetic energy E_k

of the test particle is[1] $\frac{1}{2} v^2 = \frac{1}{2} \dot{\vec{T}} \cdot \dot{\vec{T}}$, whence $dE_k = \dot{\vec{T}} \cdot \ddot{\vec{T}} \, dt$.

Now $\dot{\vec{T}}$ is the acceleration, so by Newton's law[1] $\ddot{\vec{T}}$ is also the

force, $\ddot{\vec{T}} = \text{Grad } f$. Thus $dE_k = \text{Grad } f \cdot \dot{\vec{T}} \, dt$. By (2), $dE_p =$

$-\text{Grad } f \cdot \dot{\vec{T}} \, dt$. Therefore $d(E_p + E_k) = 0$, the total energy is conserved.

By 5.2, Theorem 1, every Curl-free field is conservative.

Let $\vec{f}$ be a one-form in $\mathbb{R}^3$, and M a closed surface. Then $d(\vec{f}) = \text{Curl } \vec{f}$, and (1) becomes

$$\iint_M (\text{Curl } \vec{f} \cdot \vec{N}) \, du \, dv = \int_{\partial M} \vec{f} \cdot \vec{T} \, dt \qquad (3)$$

where $\vec{N}$ is the normal to M and $\vec{T}$ is the tangent to ∂M.

For the physical interpretation of (3) we recall from elementary mechanics that if a body rotates about an axis (which we take as the Z-axis) at ω revolutions per unit time then its angular velocity is the vector $\vec{\omega}$ pointing along the z-axis with length ω, and the linear velocity $\vec{v}$ of a point at position $\vec{r}$ in the body is

$$\vec{v} = \vec{\omega} \times \vec{r} . \qquad (4)$$

Now stand this on its head. Suppose $\vec{v}$ is the velocity field of a flowing fluid into which we insert a small paddle wheel. If the field swirls around the paddle wheel, the linear velocity of a point on its edge will be $\vec{v}$, and its angular velocity $\vec{\omega}$ will satisfy (4). In a sufficiently small neighborhood of the point we may regard $\vec{\omega}$ as constant,

---

(1) We assume unit mass.

113

and solve for it by 5.1(18):

$$\vec{\omega} = \frac{1}{2} \text{Curl } \vec{v}. \tag{5}$$

Thus Curl $\vec{v}$ measures the local swirling, or rotation, of the field $\vec{v}$, and, by (4), Curl $\vec{v}$ is perpendicular to $\vec{v}$. [(1)]

Returning to (3), we look upon $\vec{f}$

as the velocity field of a flow. Then

Curl $\vec{f}$ is its local angular velocity

vector, and   Curl $\vec{f} \cdot \vec{N}$ is the compo-

nent of this rotation which is in the

tangent plane of M.   If we add up the

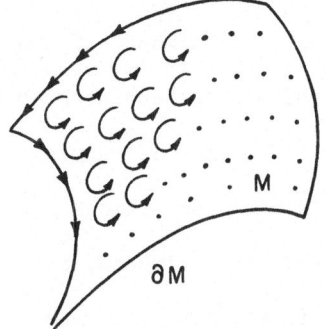

local rotations, adjacent ones cancel inside M (see Figure), leaving only the residual flow along ∂M. The latter is the right side of (3). Thus (3) is physically (or geometrically) evident by the principle of orientation cancellation.

The idea of fluid flow affords a physical confirmation that Curl Grad f = 0   for any f: if water flows down a mountain, little paddle wheels on the mountain-side will not rotate. Here f is the altitude of the mountain above each point of the base plane, Grad f is the water flow velocity field, and   Curl Grad f = 0   because the paddle wheels do not rotate.

Let $\vec{f}$ be a two-form in $\mathbb{R}^3$,   and M a solid region with closed boundary. Then $d(\vec{f}) = \text{Div } \vec{f}$ and (1) is

(1)   In 5.2, Example 2, $\vec{e}_1$ is the angular velocity of the field $\vec{g}$.
A paddle wheel in the (x_2,x_3) plane centered at $\vec{0}$ will be spun by $\vec{g}$ with angular velocity $\vec{e}_1$.

$$\iiint_{M} \text{Div } \vec{f} \; dx_1 \, dx_2 \, dx_3 \; = \; \iint_{\partial M} \vec{f} \cdot \vec{N} \; du \; dv \qquad (6)$$

where $\vec{N}$ is the normal to ∂M.

The right side is a so-called <u>flux integral</u>. If we look upon $\vec{f}$ as a field of force, this integral measures the total (signed) intensity of the field as it impinges on the "body" M. The total flux is, roughly speaking, the "number" of lines of force going out of M minus the "number" going in, taking into account also the angle at which the field lines cross ∂M and the varying magnitude of $\vec{f}(\vec{x})$. Other things equal, if the field lines diverge, as from a source (see Figure), the density (lines per unit area) on the forward side of a region M will be greater than that on the back side, so if the areas (forward and back) are comparable, $\text{Div } \vec{f} \neq 0$ in M. This is the case for the field $\vec{r}$: $\text{Div } \vec{r} = 3$ (5.2(12)). For the parallel uniform field $\vec{f}_{u} \equiv \hat{u}$, the flux in and out of any region is the same, and $\text{Div } \vec{f}_{u} = 0$ everywhere (5.2(15)).

One might suppose, from the foregoing examples, that $\text{Div } \vec{f} \neq 0$ exactly when the field lines converge or diverge to or from one or more points, but this is wrong. One must take into account also the field strength. Here is an example. Let a positive unit charge be placed at $\vec{0}$. This causes an electric field $\vec{E}$, where

$$\vec{E}(\vec{x}) \; = \; \frac{\vec{x}}{r^3} \qquad (7)$$

by Coulomb's law (we write $r = \|\vec{x}\|$; then $\dfrac{\vec{x}}{r}$ is a unit outward vector,

and the force, in appropriate units, is $\dfrac{1}{r^2}$ directed outward, hence

$\dfrac{\vec{x}}{r^3}$). By the Leibniz rule formula 5.1(12), $\operatorname{Div} \vec{E} = \operatorname{Div}(\dfrac{1}{r^3}\vec{x}) = $

$\operatorname{Grad}(\dfrac{1}{r^3}) \cdot \vec{x} + \dfrac{1}{r^3} \operatorname{Div} \vec{x};$ now $\operatorname{Grad}(r^{-3}) = -3r^{-4}\operatorname{Grad} r = -3r^{-4} \cdot \dfrac{\vec{x}}{r},$ and

$\operatorname{Div} \vec{x} = 3,$ whence

$$\operatorname{Div} \vec{E} = -\dfrac{3}{r^5} r^2 + \dfrac{3}{r^3} = 0, \tag{8}$$

even though the field lines diverge from $\vec{0}$. Note the interesting fact

that $\operatorname{Div}(\dfrac{\vec{x}}{r^n}) = 0$ if and only if $n = 3:$ $\operatorname{Grad}(r^{-n}) = -nr^{-n-1} \cdot \dfrac{\vec{x}}{r},$

whence $\operatorname{Div}(\dfrac{\vec{x}}{r^n}) = -nr^{-n-2}r^2 + r^{-n} \cdot 3 = (3-n)r^{-n}.$ That is, among the

fields $\vec{f}_n = r^{-n}\vec{x},$ the only one which is a Curl is the Coulomb

field (7). Since experiments show that electric fields give rise to

magnetic force fields, our little calculation shows that electric force

must obey an inverse-square law, and thus "predicts" Coulomb's law.

Exercise. A field $\vec{f}$ of force is called <u>central</u> if there is a given

scalar function ϕ such that $\vec{f}(\vec{x}) = \phi(r)\vec{x}.$ Show for such
a field that

$$\operatorname{Div}(\phi(r)\vec{x}) = r\phi'(r) + 3\phi(r).$$

Given scalar fields f and $g,$ by 5.1(12) we have

$$\operatorname{Div}(f \operatorname{Grad} g) = (\operatorname{Grad} f) \cdot (\operatorname{Grad} g) + f\Delta g,$$

$$\operatorname{Div}(g \operatorname{Grad} f) = (\operatorname{Grad} g) \cdot (\operatorname{Grad} f) + g\Delta f.$$

Subtracting, we find

$$\text{Div}(f \text{ Grad } g - g \text{ Grad } f) = f\Delta g - g\Delta f .$$

Therefore by the Divergence integral theorem (6) we get

$$\iiint_M (f\Delta g - g\Delta f) \, dx_1 \, dx_2 \, dx_3 =$$

(9)

$$\iint_{\partial M} (f \text{ Grad } g - g \text{ Grad } f) \cdot \vec{N} \, du \, dv .$$

This formula finds frequent application in physical arguments.

Chapter 6
Extremals

6.1 Generic Extremals

By a _generic extremal_ for a function we mean a point at which it is differentiable and takes there an extreme value. Endpoints and other points where the function in question fails to be differentiable are excluded from discussion.

Given $f : \mathbb{R}^k \to \mathbb{R}^1$, a point $\vec{a}$ such that

$$f_1(\vec{a}) \; = \; f_2(\vec{a}) \; = \; \dots \; = \; f_k(\vec{a}) \; = \; 0 \tag{1}$$

is called a _critical point_. Clearly a generic extremal must be a critical point, but the reverse need not hold, as in the case $k = 1$ $(f : \mathbb{R}^1 \to \mathbb{R}^1)$ of one-variable calculus. In one-variable calculus, if $x = a$ is a critical point, then near $x = a$ we have

$$f(x) \; = \; f(a) + \frac{(x-a)^2}{2!}\, f''(a) + \dots \tag{2}$$

(Taylor's theorem without the linear term because $f'(a) = 0$). If $f''(a) > 0$, then $f(a)$ is a minimum; if $f''(a) < 0$, then $f(a)$ is a maximum; if

$f''(a) = 0$, the procedure is inconclusive, and further terms in (2) are required to decide the case.

In k variables the procedure is the same, _mutatis mutandis_, except for a note of complexity. Given $f : \mathbb{R}^k \to \mathbb{R}^1$, if $\vec{x} = \vec{a}$ is a critical point, then near $\vec{x} = \vec{a}$ we have

$$f(\vec{x}) = f(\vec{a}) + \frac{1}{2} \sum_{i,j} f_{ij}(\vec{a})(x_i - a_i)(x_j - a_j) + \ldots . \tag{3}$$

The matrix H, where

$$H(\vec{x}) = ((f_{ij}(\vec{x}))) \tag{4}$$

is called the _Hessian_ of f at $\vec{x}$. Put $\vec{x} - \vec{a} = \vec{h}$. Then the second-order term in (3) is

$$H(\vec{a})\vec{h} \cdot \vec{h} , \tag{5}$$

and the procedure will yield a maximum or a minimum according as $H(\vec{a})\vec{h} \cdot \vec{h}$ is negative or positive (respectively) for all sufficiently small $\vec{h}$; and the procedure is inconclusive if (5) vanishes for some $\vec{h}$. The procedure is thus conclusive if (5) is of one sign for all sufficiently small $\vec{h}$. But since $H(\vec{a})$ is linear, a scale change in $\vec{h}$ will not affect the sign of (5). Therefore the procedure is conclusive if (5) is of one sign for all $\vec{h}$, and we turn to linear algebra[1] to find a criterion for this.

Since f is real, so are all its partial derivatives, and by the equality of mixed partials its Hessian is a (real) symmetric matrix.

(1) See Schreier-Sperner [5], Section 24, for the necessary material.

Therefore it may be diagonalized. Let O be the orthogonal matrix such that $OH(\vec{a})^t O = D(\vec{a})$ is diagonal,

$$D(\vec{a}) = \begin{pmatrix} \lambda_1 & & & 0 \\ & \lambda_2 & & \\ & & \ddots & \\ 0 & & & \lambda_k \end{pmatrix}, \qquad (6)$$

where $\lambda_1, \ldots, \lambda_k$ are the eigenvalues of $H(\vec{a})$. Now $H(\vec{a})\vec{h} \cdot \vec{h} = D(\vec{a})\vec{\ell} \cdot \vec{\ell}$, where $\vec{\ell} = O\vec{h}$ is arbitrary because $\vec{h}$ is, and

$$D(\vec{a})\vec{\ell} \cdot \vec{\ell} = \sum \lambda_i \ell_i^2 \qquad (7)$$

is of one sign for all $\vec{\ell}$ if and only if $\lambda_1, \lambda_2, \ldots, \lambda_k$ are all of one sign.

This general result may or may not be practical in a given problem in k variables. However, for two variables it yields immediately the standard recipe. Let $\vec{f} : \mathbb{R}^2 \to \mathbb{R}^1$ be given, and let $\vec{x} = \vec{a}$ be a critical value, $f_1(\vec{a}) = f_2(\vec{a}) = 0$. The Hessian of f is[1]

$$H = \begin{pmatrix} f_{11} & f_{12} \\ f_{21} & f_{22} \end{pmatrix} \qquad (8)$$

with $f_{21} = f_{12}$. The product of its eigenvalues is

$$\det H = f_{11}f_{22} - (f_{12})^2, \qquad (9)$$

---

[1] Here and in what follows we suppress everywhere the argument $\vec{a}$ in all functions.

and the sum of the eigenvalues is

$$\text{tr } H = f_{11} + f_{22} \, . \quad \text{(1)} \qquad\qquad (10)$$

The two eigenvalues have the same sign, therefore, if and only if $f_{11}f_{22} - (f_{12})^2 > 0,$ and this sign is positive (negative) according as $\text{tr } H = f_{11} + f_{22} > 0$ (< 0) respectively. If $\det H < 0,$ the eigenvalues have opposite signs and the point is called a <u>saddle</u>: f increases in one direction and decreases in the orthogonal direction. If $\det H = 0,$ the procedure is inconclusive.

<u>Example</u> 1. $f(\vec{x}) = \sqrt{x_1^2 + x_2^2} = r$. Then

$$f_1 = \frac{x_1}{r}, \quad f_2 = \frac{x_2}{r} \, . \qquad\qquad (11)$$

Neither of these functions is defined at $\vec{x} = \vec{0},$ the suspected extremal. In terms of the polar coordinate $r,$ f has a corner at $r = 0$. The surface is in fact an inverted cone with apex at $\vec{0}$. Thus $\vec{x} = \vec{0}$ is an extremal (a minimum), but not a generic extremal, because f is not differentiable at $\vec{x} = \vec{0}$.

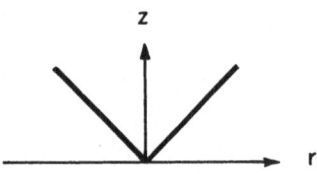

<u>Example</u> 2. $f(\vec{x}) = x_1^2 + x_2^2 = r^2$. Here

$$f_1 = 2x_1, \quad f_2 = 2x_2, \quad f_{12} = f_{21} = 0 \, ,$$
$$f_{11} = f_{22} \equiv 2 \, . \qquad\qquad (12)$$

---

(1) Recall that $\text{tr } A$ denotes the trace of the matrix A and equals the sum $\sum a_{ii}$ of its main diagonal entries.

The only critical point is $\vec{x} = \vec{0}$. At $\vec{x} = \vec{0}$ we have det $H = 4 > 0$, tr $H = 4 > 0$; hence $f(\vec{0})$ is a minimum.

Example 3. $f(\vec{x}) = x_1 x_2$. Then

$$f_1 = x_2, \quad f_2 = x_1, \quad f_{12} \equiv 1,$$

$$(13)$$

$$f_{11} = f_{22} = 0.$$

The only critical point is $\vec{x} = \vec{0}$. At $\vec{x} = \vec{0}$ we have det $H = -1$, tr $H = 0$. Therefore the eigenvalues are $\lambda_{1,2} = \pm 1$, and $\vec{x} = \vec{0}$ is a saddle point.

Example 4. $f(\vec{x}) = x_1^2 x_2^3$. Then

$$f_1 = 2x_1 x_2^3, \quad f_2 = 3x_1^2 x_2^2, \quad f_{12} = 6x_1 x_2^2,$$

$$(14)$$

$$f_{11} = 2x_2^3, \quad f_{22} = 6x_1^2 x_2,$$

we have Det $H = 12x_1^2 x_2^4 - 36x_1^2 x_2^4 = -24x_1^2 x_2^4 \leq 0$; and tr $H = 2x_2^3 + 6x_1^2 x_2 = 2x_2(x_2^2 + 3x_1^2)$, whence the sign of tr H is the sign of x_2. The x_1 and x_2 axes are both lines of critical points; on both lines det H vanishes; on the x_1 axis both eigenvalues vanish; on the x_2-axis, one eigenvalue is non-zero, having the sign of x_2. The graph of f is a vigorous everywhere differentiable landscape of descending valleys and ascending ridges, having no maxima, no minima, and no saddles.

6.2 Extremals under Constraints

A constraint in $\mathbb{R}^k$ is a functional relation on the variables $x_1, x_2, \ldots, x_k$; it may be expressed in terms of a function $\phi : \mathbb{R}^k \rightarrow \mathbb{R}^1$ in the form

$$\phi(\vec{x}) = 0. \qquad (1)$$

The extremal problem for a given function $f : \mathbb{R}^k \rightarrow \mathbb{R}^1$ under the constraint (1) is this: find the extremes among the set of numbers $f(\vec{x})$ where $\vec{x}$ is "constrained" to satisfy (1). In geometric terms this means: regard f as a function from the variety $Z(\phi)$ to $\mathbb{R}^1$, $f : Z(\phi) \rightarrow \mathbb{R}^1$, and find the extreme values of this (restricted, or constrained) new function.

Example. What is the largest product which may be formed with two numbers whose sum is 10? Here the given function is $f(\vec{x}) = x_1 x_2$; the constraint is $\phi(\vec{x}) = x_1 + x_2 - 10 = 0$; $Z(\phi)$ is a straight line in the plane (see Figure); and to solve the problem we want that point on the line whose coordinates have maximum product.

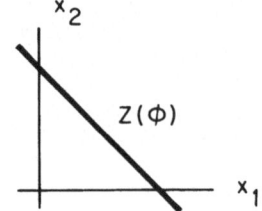

The solution to the general problem of extremals under a constraint turns on the correct understanding of "critical point" in this context. A point $\vec{a}$ in $Z(\phi)$ is critical for f under the constraint (1) if the derivatives of f in the directions of the manifold $Z(\phi)$ vanish; that is to say, if

$$D_{\vec{u}} f(\vec{a}) = 0, \quad \text{all } \vec{u} \text{ tangent to } Z(\phi), \qquad (2)$$

where $D_{\vec{u}}$ is the directional derivative 5.1(18). Note that (2) does

not entail $f_1(\vec{a}) = \ldots = f_k(\vec{a}) = 0$; a point may be critical for f

under the constraint without being critical for f in the ordinary sense

6.1(1). The foregoing Example illustrates this point.

By 5.1(20) a point $\vec{a}$ in $Z(\phi)$ is critical for f under the

constraint (1) if and only if Grad $f(\vec{a}) \cdot \vec{u} = 0$ for all tangents $\vec{u}$

to $Z(\phi)$; that is to say, if and only if Grad $f(\vec{a})$ is normal to $Z(\phi)$.

We know that Grad ϕ is normal to $Z(\phi)$.[1] We conclude that $\vec{a}$ in

$Z(\phi)$ is critical for f under the constraint (1) if and only if Grad f

and Grad ϕ are parallel at $\vec{a}$. In components this is

$$f_i(\vec{a}) = \lambda \phi_i(\vec{a}), \quad i = 1, 2, \ldots, k, \tag{3}$$

for some constant λ. This suggests the following method, due to

Lagrange and called the method of Lagrange's multiplier, for finding the

critical points $\vec{a}$: Form the auxiliary function

$$F(\vec{x}, \lambda) = f(\vec{x}) - \lambda \phi(\vec{x}) \tag{4}$$

where λ, the "multiplier", is simply postulated; extremize F in the

ordinary way as a function of k+1 variables $x_1, x_2, \ldots, x_k, \lambda$; and

eliminate λ. The method works because the equations $F_i = 0$, $i = 1, 2, \ldots, k$

are the conditions (3) and the equation $F_\lambda = 0$ is the constraint (1).

Example (Continued). The auxiliary function is

---

(1) See 3.3(15). One need only specialize the discussion of 3.3(9) et
seq. to a single function, and note that the vector $\vec{\phi}^\mu$ of 3.3(11)
is what we would now call Grad ϕ^μ.

$$F(\vec{x},\lambda) = x_1 x_2 - \lambda(x_1 + x_2 - 10) .$$

Now $F_1 = x_2 - \lambda$, $F_2 = x_1 - \lambda$, $F_3 = x_1 + x_2 - 10$; setting these equal to 0 we have $x_1 = \lambda = x_2$, which eliminates λ; and putting $x_1 = x_2$ in the constraint equation $x_1 + x_2 - 10 = 0$ we find $x_1 = x_2 = 5$, so the answer is 25.

The method extends at once to extremal problems with several constraints. One introduces a multiplier for each constraint and proceeds as before, relying upon the discussion of tangents and normals to varieties of the form $Z(\phi^1, \phi^2, \ldots)$ as given in 3.3. We omit the details.

The method of Lagrange's multiplier yields an elegant proof of the important __theorem on the geometric and arithmetic mean__, which states that

$$(a_1 a_2 \ldots a_n)^{\frac{1}{n}} \leq \frac{1}{n}(a_1 + \ldots + a_n) \tag{5}$$

for any $a_i > 0$, $i = 1, 2, \ldots, n$, and all $n = 1, 2, \ldots$. Put, for $\vec{x}$ in $\mathbb{R}^n$, $x_i > 0$,

$$f(\vec{x}) = (\Pi x_i)^{\frac{1}{n}}, \tag{6}$$

$$\phi(\vec{x}) = \sum x_i - 1, \tag{7}$$

and form the auxiliary function

$$F(\vec{x},\lambda) = f(\vec{x}) - \lambda\phi(\vec{x}) .$$

Now $\quad f_i = \frac{1}{n}(\Pi x_i)^{\frac{1}{n}-1} \underset{j\neq i}{\Pi} x_j = \frac{1}{n} \underset{\ell\neq i}{\Pi} x_\ell^{\frac{1}{n}} \cdot x_i^{\frac{1}{n}-1} = \frac{1}{n} f(\vec{x}) \cdot x_i^{-1}, \quad$ whence

$$F_i(\vec{x}) = \frac{1}{n} f(\vec{x}) \cdot x_i^{-1} - \lambda .$$

Therefore the equations $F_i = 0$ eliminate λ and, at the same time, show that $x_1 = x_2 = \ldots = x_n$ at a constrained extremal. The constraint (7) shows that in fact $x_1 = \ldots = x_n = \frac{1}{n}$ at such points, so that the constrained extreme value is $(\Pi \frac{1}{n})^{\frac{1}{n}} = \frac{1}{n}$. Therefore $f(\vec{y}) \leq \frac{1}{n}$ for all $\vec{y} \in Z(\phi)$. For every $\vec{x}$ with $\sum x_i \neq 0$ we have $\vec{y} = \frac{1}{c}\vec{x} \in Z(\phi)$, where $c = \sum x_i$. Now observe that f is homogeneous: $f(t\vec{x}) = (t^n \Pi x_i)^{\frac{1}{n}} = tf(\vec{x})$. Hence for any $\vec{x}$ with $c = \sum x_i \neq 0$, $\frac{1}{c} f(\vec{x}) = f(\frac{1}{c}\vec{x}) = f(\vec{y}) \leq \frac{1}{n}$, whence $f(\vec{x}) \leq \frac{1}{n}\sum x_i$. Since the theorem requires $x_i > 0$, whence $c \neq 0$, the proof is complete.

Chapter 7
Integral geometry

7.1 Measure of Points and Lines in $\mathbb{R}^2$

Let $\mathbb{E}$ denote the set of all rigid motions of the plane.[1]   A
rigid motion T consists of a rotation $R(\alpha)$ through an angle α
followed by a translation by a vector $\vec{a}$. In symbols,

$$T\vec{x} = \vec{a} + R(\alpha)\vec{x} .\tag{1}$$

Putting as usual $\vec{y} = T\vec{x}$ we have for (1) in components

$$y_1 = a_1 + x_1 \cos\alpha - x_2\sin\alpha ,$$

$$\tag{2}$$

$$y_2 = a_2 + x_1 \sin\alpha + x_2\cos\alpha .$$

If $f(\vec{x}) \geq 0$ for all $\vec{x} \in \mathbb{R}^2$ the formula

$$m(E) = \int_E f(\vec{x})\,dx_1 \wedge dx_2\tag{3}$$

----

(1) See Schreier-Sperner [5], Section 12, especially pp. 163–168. It is
there proved that a rigid motion consists of a rotation together
with a translation.

defines a <u>measure</u>, or method of assigning areas, for subsets $E \subset \mathbb{R}^2$.
The customary area is given by (3) with $f \equiv 1$. The customary area is
<u>invariant under the action of</u> $\mathbb{E}$ in the sense that

$$m(TE) = m(E) \tag{4}$$

for every $T \varepsilon \mathbb{E}$. We now seek to characterize those measures (3) which
have this property. By the change of variable formula 3.1(8) we have

$$m(TE) = \int_E (f \circ T)(\vec{y}) \frac{\partial(x_1, x_2)}{\partial(y_1, y_2)} dy_1 \wedge dy_2 .$$

Looking to (2) we find $\dfrac{\partial(x_1, x_2)}{\partial(y_1, y_2)} = 1$ (using 3.1(23)) whence we may
write

$$m(TE) = \int_E (f \circ T)(\vec{x}) dx_1 \wedge dx_2 .$$

If m is invariant under $\mathbb{E}$, then by (4) we have $\int_E (f \circ T)(\vec{x}) dx_1 \wedge dx_2 = \int_E f(\vec{x}) dx_1 \wedge dx_2$, this for every E; whence $f \circ T = f$ for every T,
so that finally f must be constant. Thus the only measure of the form
(3) invariant under $\mathbb{E}$ is, apart from a constant factor, the customary
measure. This is called the <u>measure of points</u>, since it measures sets
of points in $\mathbb{R}^2$, and invariantly with respect to $\mathbb{E}$. It is denoted
dP. Thus

$$dP = dx_1 \wedge dx_2 . \tag{5}$$

By a very similar process we can construct the analog of this for straight lines in the plane: a measure of sets of straight lines which is invariant under the action of $\mathbb{E}$. We express each straight line L in its polar form

$$L : x_1 \cos \phi + x_2 \sin \phi = p$$

with notation as in the figure. Then each L is specified uniquely by a pair (p, ϕ) of real numbers with $0 \leq p < \infty$, $0 \leq \phi < 2\pi$. Every $T \in \mathbb{E}$ carries straight lines to straight lines, of course, and if the polar distance and angle of $T(L)$ are (q, θ), where T is given by (2), then we find

$$q = p - a_1 \cos \alpha - a_2 \sin \alpha \,,$$

$$\theta = \phi - \alpha \,.$$

(6)

If $m(E) = \int_E f(p, \phi) \, dp \wedge d\phi$ is a proposed measure for an arbitrary set E of straight lines, then by the change of variable formula we have $m(TE) = \int_E (f \circ T)(q, \theta) \frac{\partial(p, \phi)}{\partial(q, \theta)} \, dq \wedge d\theta$; we find $\frac{\partial(p, \phi)}{\partial(q, \theta)} = 1$ by (6); and if m is assumed to be invariant under the action (6) of $\mathbb{E}$ then we conclude as before that f is constant. We choose the constant 1 and denote by dL the resulting measure of lines,

$$dL = dp \wedge d\phi \,.$$

(7)

Here is an application of the measure of lines. Let a smooth curve C

be given parametrically in terms of

its arc length s .   Each line L

which intersects C is determined

by the value of s determining the

point $(x_1(s), x_2(s))$ of intersection

and the angle θ between the tangent

to C and L (see Figure). Let us

express dL in terms of s and θ .

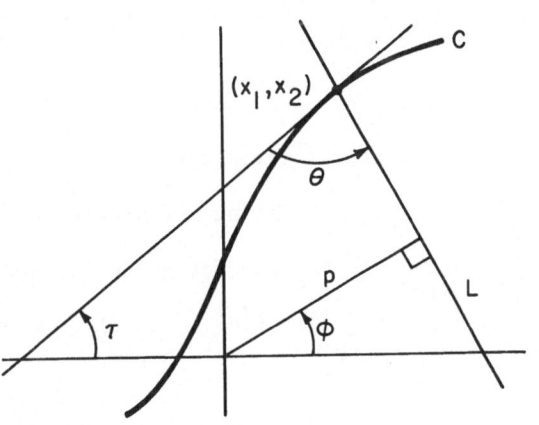

Since (x_1, x_2) is on L we have $p = x_1 \cos \phi + x_2 \sin \phi$.   Hence

$dp = (\cos \phi \, dx_1 + \sin \phi \, dx_2) + (-x_1 \sin \phi + x_2 \cos \phi) d\phi$.   Now $dx_1 = \cos \tau \, ds$

and $dx_2 = \sin \tau \, ds$, where τ is the angle between the x_1 axis and the

tangent to C .   Hence $dp = \cos(\phi - \tau) ds + (-x_1 \sin \phi + x_2 \cos \phi) d\phi$.   Taking

wedge product with $d\phi$ on the right we then have $dp \wedge d\phi = \cos(\phi - \tau) ds \wedge d\phi$.

Adding angles in the large triangle of the figure we have $\tau + \theta + \pi/2 - \phi = \pi$,

or

$$\phi = \tau + \theta - \pi/2 .$$

Therefore $d\phi = \tau' ds + d\theta$, $ds \wedge d\phi = ds \wedge d\theta$,   and $\cos(\phi - \tau) =$

$\cos(\theta - \pi/2) = -\sin(\theta)$. Since areas (measures) must be non-negative, we

take absolute value here and have

$$dL = |\sin \theta| ds \wedge d\theta . \qquad (8)$$

Using this formulation of dL we may easily compute

$$\int_{L \cap C \neq \emptyset} \#(L \cap C)\,dL\,,$$

where $\#(L \cap C)$ denotes the number of points in which L intersects C.
The integral gives the line measure of the set of lines which intersect
C, each line being weighted according to the number of its intersections
with C . By the foregoing discussion the integral equals

$$\int_0^{\ell(C)} ds \cdot \int_0^\pi |\text{Sin } \theta|\,d\theta \;=\; 2\ell(C)\,, \qquad \text{where} \quad \ell(C) \quad \text{denotes the length of}\quad C.$$

Thus

$$\int_{L \cap C \neq \emptyset} \#(L \cap C)\,dL \;=\; 2\ell(C)\,. \qquad (9)$$

If C is a closed convex curve,[1] then $\#(L \cap C) = 2$ for all L which
meet C non-tangentially,[2] whence (in abbreviated notation) $\int \#(L \cap C)\,dL =$
$\ell(C)$ in that case.

7.2 Kinematic Measure

Let K be a plane geometric figure such as a line segment, or a
curve segment, or a bounded region with a piece-wise smooth[3] simple
closed boundary curve. Figures of the latter type are called domains.
Figures of infinite extent (which is to say, unbounded figures) such as

---

(1) A curve C is convex if the straight line determined by every pair
of points of C has no further intersections with C .

(2) The set of lines tangent to C has line measure 0 because, for such
lines, ϕ and p are functions of s only, so that $dp \wedge d\phi$ will
involve $ds \wedge ds = 0$, which is to say, $dL = 0$.

(3) This means it is differentiable (and so has a tangent) at all except
a finite number of its points, at which it turns sharply. Such points
are called corners.

straight lines are excluded. We wish to measure sets of figures congruent

to K, or what is the same, sets of positions of K .   Such a measure

would be an analog for general figures of the measure of lines. To

describe the possible positions of K we think of K as mobile and

endowed with a reference frame attached rigidly to it. Each position of K

may then be given by specifying the coordinates (x_1,x_2) in the fixed

reference frame of the mobile origin and the angle ϕ between the fixed

and mobile x_1 axes. K has in this way the <u>three</u> coordinates (x_1,x_2,ϕ).

Let us apply to K a rigid motion $T \in \mathbb{E}$, with angle α and vector

$\vec{a}$, as in 7.1(2). If the coordinates of TK are (y_1,y_2,θ) then

clearly

$$y_1 = a_1 + x_1 \cos \alpha - x_2 \sin \alpha ,$$

$$y_2 = a_2 + x_1 \sin \alpha + x_2 \cos \alpha , \tag{1}$$

$$\theta = \phi + \alpha.$$

To find the most general measure of sets of figures congruent to K which

is invariant under the action of $\mathbb{E}$, we proceed just as we did for points

and lines. Put $m(E) = \int_E f(x_1,x_2,\phi)\,dx_1 \wedge dx_2 \wedge d\phi$, where E is such a

set and $f \geq 0$; whence $m(TE) = \int_E (f \circ T)(y_1,y_2,\theta)\frac{\partial(x_1,x_2,\phi)}{\partial(y_1,y_2,\theta)}\,dy_1 \wedge dy_2 \wedge d\theta$;

observe that $\frac{\partial(x_1,x_2,\phi)}{\partial(y_1,y_2,\theta)} = 1$; and conclude as before that f is constant.

Choosing 1 for this constant we have an invariant measure for congruent

figures K which we denote by dK ,

$$dK = dx_1 \wedge dx_2 \wedge d\phi, \tag{2}$$

and refer to as the <u>kinematic-measure</u>, for the obvious reason. We emphasize that the kinematic measure applies at once to every figure K, and so

constitutes a remarkable generalization of the measure of lines.

Let a figure K range over a set E of positions, the kinematic

measure of which is then $\int_E dK$, by definition. If we view this situation

from the mobile reference frame, letting the world turn under the worm,

so to speak, and ask for the kinematic measure of the set of positions of

the fixed reference frame, the answer is again $\int_E dK$. For if (x_1,x_2,ϕ)

are the coordinates and angle in the fixed frame of the mobile frame, then

the coordinates and angle (x_1',x_2',ϕ') of the fixed frame in the mobile

frame are

$$x_1' = -x_1 \cos \phi - x_2 \sin \phi ,$$

$$x_2' = x_1 \sin \phi - x_2 \cos \phi ,$$

$$\phi' = -\phi ;$$

whence $\dfrac{\partial(x_1',x_2',\phi')}{\partial(x_1,x_2,\phi)} = -1$, so that, taking absolute value because

measure must be non-negative, we find that the change of variable caused

by interchanging the reference frames does not change the measure in

question. This property of the kinematic measure is known as _invariance_

under inversion of the motion. It is an interesting property, as the

following immediate application shows. Suppose we require the kinematic

measure of all congruent domains K containing a fixed point P, in

symbols $\int_{P\in K} dK$. By inversion this is the same as fixing K and letting

P range over its interior, which yields $\int_{P\in K} dx_1 \wedge dx_2 \cdot \int_0^{2\pi} d\phi = 2\pi \cdot a(K)$,

where a(K) denotes the area of K . Thus

$$\int_{P \epsilon K} dK = 2\pi \cdot a(K) . \tag{3}$$

7.3 Formulae of Poincaré and Blaschke

Formula 7.1(9) has an analog in which lines are replaced by smooth curves. Let C_o be a fixed curve parametrized as $(x_1(s), x_2(s))$ by its arc length s. Denote its length $\ell(C_o)$ by ℓ_o. Let C denote a mobile curve of length $\ell(C) = \ell$. Denote the mobile axes as X_1, X_2, let the mobile origin have coordinates (y_1, y_2) relative to the fixed frame, and let C be given parametrically relative to the fixed frame as $(X_1(\sigma), X_2(\sigma))$, where σ is its arc length. Then relative to the fixed frame the equations of C are

$$x_1 = y_1 + X_1(\sigma) \cos \phi - X_2(\sigma) \sin \phi ,$$

$$x_2 = y_2 + X_1(\sigma) \sin \phi + X_2(\sigma) \cos \phi ,$$

where ϕ is the angle between the x_1 and X_1 axes. We seek an expression for the kinematic measure $dC = dy_1 \wedge dy_2 \wedge d\phi$ of C in terms of s, σ, and the angle ω between the tangents to C_o and C at a point of intersection. Their points of intersection are given in terms of s and σ by

$$x_1(s) = y_1 + X_1(\sigma) \cos \phi - X_2(\sigma) \sin \phi ,$$

$$x_2(s) = y_2 + X_1(\sigma) \sin \phi + X_2(\sigma) \cos \phi .$$

Therefore $dx_1 = dy_1 + dX_1 \cos \phi - dX_2 \sin \phi - (X_1 \sin \phi + X_2 \cos \phi) d\phi$, and dx_2 similarly. Taking account of the relations

$$dx_1 = \cos \theta_o \, ds, \qquad dx_2 = \sin \theta_o \, ds,$$

$$dX_1 = \cos \theta \, d\sigma, \qquad dX_2 = \sin \theta \, d\sigma,$$

where θ_o is the angle between the x_1 axis and the tangent to C_o at an intersection point, and θ is the angle between the X_1 axis and the tangent to C at the same intersection point, we find after trivial rearrangements

$$dy_1 = \cos \theta_o \, ds - \cos(\theta+\phi) \, d\sigma + (X_1 \sin \phi + X_2 \cos \phi) d\phi,$$

$$dy_2 = \sin \theta_o \, ds - \sin(\theta+\phi) \, d\sigma - (X_1 \cos \phi - X_2 \sin \phi) d\phi;$$

whence, by the algebra of the wedge product, we have at once $dy_1 \wedge dy_2 \wedge d\phi = -\cos \theta_o \sin(\theta+\phi) \, ds \wedge d\sigma \wedge d\phi + \sin \theta_o \cos(\theta+\phi) \, ds \wedge d\sigma \wedge d\phi$, or

$$dC = |\sin(\theta_o-\theta-\phi)| \, ds \wedge d\sigma \wedge d\phi . \qquad (1)$$

Now $\theta+\phi$ is the angle between C and the x_1-axis, whence $\theta_o - (\theta+\phi)$ is the angle between the tangents to C and C_o at their intersection point. This angle we have called ω; hence $\omega = \theta_o - \theta - \phi$; and since θ_o, θ are functions of s, σ only, we have from (1) the expression

$$dC = |\sin \omega| \, ds \wedge d\sigma \wedge d\omega \qquad (2)$$

of dC in terms of s, σ, and ω, which was our goal.

135

Applying (2) just as we did 7.1(8) we have at once

$$\int \#(C \cap C_o) dC = 4\ell\ell_o,$$ (3)

where $\#(C \cap C_o)$ is the number of intersections of C and C_o. This is known as **Poincaré's formula**. It is quite analogous to 7.1(9). The factor 4 arises here because in the present circumstance the angle ω must be permitted the full variation $0 \le \omega \le 2\pi$.

Put $\#(C \cap C_o) = n(C)$ and let ω_i, $i = 1, 2, \ldots, n(C)$, denote the angles between C and C_o at the $n(C)$ intersection points. At each intersection we calculate $\int \omega \, dC$. Using (2), and putting $0 \le \omega < \pi$, we have $\int \omega \, dC = 2 \int_0^\pi \omega \sin \omega \cdot \int_0^{\ell_o} ds \int_0^\ell d\sigma = 2\pi\ell\ell_o$. Summing over all intersections we have the variant

$$\int \left(\sum_{i=1}^{n(C)} \omega_i \right) dC = 2\pi\ell\ell_o$$ (4)

of (3) which we shall need in a moment.

Let K_o be a convex domain[1] with smooth boundary C_o. Denote the area of K_o by $a(K_o) = a$, and the length of C_o by ℓ. If τ is the angle between the tangent to C_o and the x_1 axis and s is arc length on C_o, then for the ordinary line integral $\int_{C_o} \tau'(s) ds$ we have the value 2π because C_o is (by assumption) a simple closed curve.

---

(1) K is convex by definition when the line segment joining any two points of K lies entirely in K.

Let K be a mobile domain congruent
to K_o. Let C be its boundary, and
let t denote arc length on C . Let
P(s), Q(t) be generic points on C_o, C
respectively. If $K \cap K_o \neq \emptyset$, then

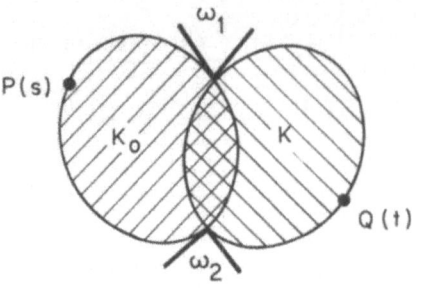

$$\int_{P(s)\in K} \tau'(s)\,ds + \int_{Q(t)\in K_o} \tau'(t)\,dt + \sum \omega_i = 2\pi , \qquad (5)$$

where ω_i are the external angles at the points of intersection of C
and C_o, because the left side counts the total turning (including its
jumps at the corners) of the tangent to the boundary of $K \cap K_o$ as a
point traverses it once, and this boundary is itself a simple closed
curve.

Consider now

$$I = \int_{P(s)\in K} \tau'(s)\,ds\,dK .$$

On one hand $I = \int_{K_o} \tau'(s)\,ds \cdot \int_{P(s)\in K} dK ;$ the first factor is 2π (as
noted above), and the second equals $2\pi a$, by 7.2(3). On the other hand,
$I = \int_{K \cap K_o \neq 0} dK \cdot \int_{P(s)\in K} \tau'(s)\,ds = \int\{ \int_{P(s)\in K} \tau'(s)\,ds\}dK .$ Putting together
these results we have

$$\int\{ \int_{P(s)\in K} \tau'(s)\,ds\}dK = 4\pi^2 a . \qquad (6)$$

By reversing the roles of K_o and K, and then putting $dK_o = dK$ by the invariance of the kinematic measure under inversion of the motion, we have also

$$\int \left\{ \int_{Q(t) \epsilon K_o} \tau'(t) \, dt \right\} dK = 4\pi^2 a^2 . \tag{7}$$

Integrating both sides of (5) with respect to the kinematic-measure dK, we get $4\pi^2 a$ for each of the first two terms on the left side, by (6) and (7); for the last term we invoke (4) with $\ell = \ell_o$ (because C and C_o are congruent), and, noting that $dC = dK$ (because the positions of K and of its boundary are of course mutually determinate), we get $2\pi \ell^2$; and the right side is $2\pi \int_{K \cap K_o \neq \emptyset} dK$. Assembling the several pieces we have finally

$$\int_{K \cap K_o \neq \emptyset} dK = 4\pi a + \ell^2 . \tag{8}$$

This is a special case of a result known as the _fundamental formula of Blaschke_.[1] It is an interesting extension of 7.2(3).

The formula of Blaschke (8) also yields, together with Poincaré's formula (3), a proof of the so-called _isoperimetric inequality_, which states, in the present context[2] (that is, for a convex domain K of area $a(K) = a$ with smooth boundary of length ℓ) that

---

(1) An account of the general result of Blaschke is given in Santalo [4], page 34, from which the foregoing special case has been adapted.

(2) The result holds in greater generality for convex curves (Santalo [4], p. 37). There is another version in which smoothness replaces convexity (see, e.g., Courant [1], volume II, p. 214).

$$\ell^2 - 4\pi a \geq 0 .$$
(9)

For the proof, we put $n(C) = \#(C \cap C_o)$ and, because $dC = dK$, we have

$$\int_{K \cap K_o \neq \emptyset} n(C) \, dK = 4\pi^2$$
(10)

by (3). We define the sequence $M_1, M_2, \ldots$ by

$$M_n = \int_{n(C) = n} dK , \qquad n = 1, 2, \ldots .$$
(11)

Observe that $M_n = 0$ if n is odd, for if $n(C) = 2j+1$ then C must have a point of tangency with C_o, and in the set of such positions (one tangency and j pairs of other intersections) the polar angle ϕ will be a function of the coordinates (x_1, x_2) of the mobile origin, whence dK will involve the vanishing form $dx_1 \wedge dx_1$; which is to say, the kinematic measure of the set in question is 0, as asserted. Putting (10) and (11) together, and taking into account what has just been shown, we have

$$4\ell^2 = 2M_2 + 4M_4 + 6M_6 + \ldots .$$

Blaschke's formula (8) gives

$$4\pi a + \ell^2 = M_2 + M_4 + M_6 + \ldots .$$

The two together entail $2\ell^2 - 8\pi a = 2M_4 + 4M_6 + \ldots ,$ or

$$\ell^2 - 4\pi a = M_4 + 2M_6 + 3M_8 + \ldots , \qquad\qquad (12)$$

which yields (9) because the right side of (12) is non-negative. Notice that if K is a circle then $0 = M_4 = M_6 = \ldots$, consistent with the elementary fact that $\ell^2 = 4\pi a$ for a circle. The converse assertion, that if $M_n = 0$ for $n \neq 2$ then K is a circle, is correct, but that is less elementary, and we shall not go into it here.

Appendix

1. The Volume Element on a Manifold

As we have argued in 3.4, the volume element on a manifold $M \subseteq \mathbb{R}^k$ of dimension $r \leq k$, given by $\vec{\psi} : \mathbb{R}^r \to \mathbb{R}^k$ say, is determined by finding the r-dimensional volume of the parallelopiped defined by the explicit basis $\vec{\psi}_j$, $j = 1, 2, \ldots, r$, of the tangent space to M. Using the law of cosines we found the formula 3.4(5) for the volume element of a surface $(r = 2)$, and we asserted the general formula in 3.4(6). We shall here establish the latter. Unfortunately the demonstration is lengthy and not entirely elementary. However, since our main applications of integration (the integral theorems of 5.3) occur in $\mathbb{R}^3$, the general formula and the ensuing long argument may be treated as optional material.

Evidently the problem may be given the following purely geometric statement: find the r-dimensional volume of the parallelopiped determined by r vectors $u_i \in \mathbb{R}^k$, $i = 1, 2, \ldots, r$, $r \leq k$; and the assertion is that the square of this volume is the sum of the squares of the $\binom{k}{r}$ minors of the r-by-k matrix whose rows are the vectors u_i. Put this way, the assertion appears as a generalization of the Pythagorean theorem. For $r = 1$ it is exactly the k-dimensional form of that theorem: the parallelopiped is a single vector, its "volume" is the length of that vector, and the square of this length is the sum of the squares of the components of

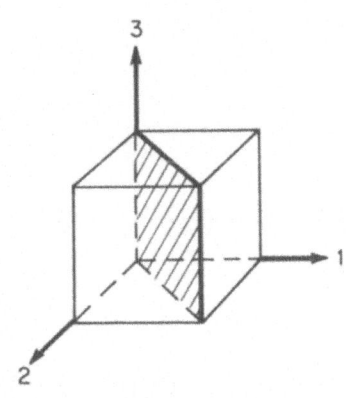

the vector, by the Pythagorean theorem.
Look next at the 2-dimensional figure
in $\mathbb{R}^3$ determined by $u_1 = (1,1,0)$ and
$u_2 = (0,0,1)$. It is the shaded square of
the figure. The perpendicular projec-
tions (or shadows) cast upon the three
coordinate planes by this figure are: unit squares on $(1,3)$ and $(2,3)$,
and a line on $(1,2)$, as one sees in the figure. The 2-dimensional
volumes (areas) of these shadows are respectively: 1, 1, and 0; the area
of the figure itself is $1 \cdot \sqrt{2} = \sqrt{2}$ (since one of its sides is the
diagonal of a unit square); and $1^2 + 1^2 = (\sqrt{2})^2$. That is, the sum of
the squares of the areas of the shadows of the figure is the square of
the area of the figure. This example displays the general scheme: there
are $\binom{k}{r}$ "coordinate r-planes", so to speak, upon which the r-dimensional
parallelopiped determined by r vectors in $\mathbb{R}^k$ casts shadows, and the
assertion is that the r-dimensional volumes of these shadows are
"Pythagorean components" of the volume of the parallelopiped.

For the demonstration we need a bit of notation. Let $[u] =$
$(u_1, u_2, \ldots, u_r)$ denote the set of vectors in $\mathbb{R}^k$ to be considered, their
components being indicated by superscripts; let $(i) = (i_1, i_2, \ldots, i_r)$
denote a set of indices, $1 \leq i_1 < \ldots < i_r \leq k$; let

$$M_r^{(i)}[u] = \det((u_m^{i_n})), \tag{1}$$

$m = 1, 2, \ldots, r$ and $i_n \in (i)$, denote the r-by-r minor got by choosing the
$i_1, i_2, \ldots, i_r$ columns from the r-by-k matrix of the u_i; let

$$M_r[u] = \{M_r^{(i)}[u]\} \tag{2}$$

denote the sequence of the $\binom{k}{r}$ minors (1) arranged in their lexicographic order; let

$$(M_r[u], M_r[v]) = \sum_{(i)} M_r^{(i)}[u] \cdot M_r^{(i)}[v] \tag{3}$$

denote the inner product in the space of $\binom{k}{r}$-tuples; finally, let $P[u]$ denote the parallelopiped determined by the u_i, and $V_r[u]$ its r-dimensional volume. In these terms our assertion is:

$$V_r[u] = \|M_r[u]\| . \tag{4}$$

If A is an orthogonal k-by-k matrix, we may apply it to $u_1, u_2, \ldots, u_r$, and we denote the resulting set $Au_1, Au_2, \ldots, Au_r$ by $[Au]$. We then have a new sequence $M_r[Au]$ of minors. It is plausible that the inner product (3) is preserved under this action of A,

$$(M_r[Au], M_r[Av]) = (M_r[u], M_r[v]), \tag{5}$$

a fact we shall use without proof.

We proceed by induction. We know (4) holds for $r = 1$ and $r = 2$. Suppose it holds for all $r \le s < k$. Given s+1 vectors $u_1, u_2, \ldots, u_s, w$, by a process of successive orthogonalization [(1)] we can express w as

---

(1) See Schreier-Sperner [5], pp. 140-141. In recent texts this construction is called the Gram-Schmidt process.

$w = a + b$, where $b \in P(u_1, u_2, \ldots, u_s)$ and a is orthogonal to u_i, $i = 1, 2, \ldots, s$. We may (and do) assume that $u_1, u_2, \ldots, u_s$ are linearly independent. It then follows that b is dependent upon them, so that $M_{s+1}^{(j)}(u_1, \ldots, u_s, b) = 0$ for every $(j) = (j_1, j_2, \ldots, j_{s+1})$, $1 \le j_1 < \ldots < j_{s+1} \le k$, whence

$$M_{s+1}^{(j)}(u_1, u_2, \ldots, u_s, w) = M_{s+1}^{(j)}(u_1, u_2, \ldots, u_s, a) , \qquad (6)$$

by the multilinearity of determinants. For the volumes we have the corresponding reduction

$$V_{s+1}(u_1, \ldots, u_s, w) = V_{s+1}(u_1, \ldots, u_s, a) , \qquad (7)$$

by the meaning of volume. Since a is orthogonal to u_i, $i = 1, 2, \ldots, s$, we have $V_{s+1}(u_1, \ldots, u_s, a) = V_s(u_1, \ldots, u_s) \cdot \|a\|_k$, where $\|a\|_k$ denotes the length of a in $\mathbb{R}^k$. By the inductive hypothesis we now have

$$V_{s+1}(u_1, \ldots, u_s, w) = \|M_s[u]\| \cdot \|a\|_k . \qquad (8)$$

Let A be an orthogonal transformation such that $Aa = (\lambda, 0, 0, \ldots, 0)$, where $\lambda = \|a\|_k$. Then, because A is orthogonal, Aa is orthogonal to $Au_1, Au_2, \ldots, Au_s$. Now $V_{s+1}(u_1, u_2, \ldots, u_s, w) = V_{s+1}(Au_1, Au_2, \ldots, Au_s, Aw)$ by the meaning of volume. Since Au_i, Aw are exactly as general as u_i, w we have, by (8),

$$V_{s+1}(Au_1, \ldots, Au_s, Aw) = \|M_s[Au]\| \cdot \|Aa\|_k . \qquad (9)$$

Since all Au_i have vanishing first coefficients (by virtue of being orthogonal to $Aa = (\lambda,0,\ldots,0))$, the (non-vanishing) minors of the $(s+1)$-by-k matrix of the Au_i and Aa consist of the (non-vanishing) minors of the s-by-k matrix of the Au_i multiplied by $\lambda = \|Aa\|_k$. Therefore the right side of (9) equals $\|M_{s+1}(Au_1,Au_2,\ldots,Au_s,Aa)\|$. At this point we invoke the invariance (5), by which we may remove A from the last expression. Then, taking account of (6), and retracing our steps back to (7), we have shown that

$$V_{s+1}(u_1,\ldots,u_s,w) = \|M_{s+1}(u_1,\ldots,u_s,w)\|,$$

which establishes the assertion QED.

2. The Algebra of Forms

The algebra of forms rests upon the axioms (1), (5), and (6) of 2.3. These assert respectively the anti-commutation of the wedge product, its homogeneity with respect to $\mathfrak{S}$ multiplication, and its distributivity with respect to form addition. To show these axioms consistent it suffices to establish this in the particular case of $\mathbb{R}^3$. In 5.1(6) we have shown how to put forms and vector fields into one-to-one correspondence in such way that the wedge and vector products correspond. Since the vector product has the properties corresponding to the axioms in question, the latter are consistent.

3. **A Remark on $Curl^2$**

Among the higher differentiations (5.2) the only ones which carry scalar fields to scalar fields are the powers Δ^n of Δ (5.2(7)). These may be applied componentwise to a vector field, and this yields new differentiations carrying vector fields to vector fields. These new differentiations may then be combined with the standard vector differentiations $Curl^k$, $Grad\ \Delta^\ell\ Div$ (5.2(7)), with k, ℓ so that the orders match. The lowest-order such combination is noteworthy:

$$Curl^2 \vec{f} = Grad\ Div\ \vec{f} - \Delta\vec{f}. \tag{1}$$

The verification is straightforward: we interprete $\vec{f}$ formwise as needed to carry out the operations; then interprete the results as vector fields; and compare components. For the left side we start with $\vec{f} = \sum a^i dx_i$. Then

$$Curl\ \vec{f} = \sum A^i dx_i ,$$

$$A^1 = a^3_2 - a^2_3, \qquad A^2 = a^1_3 - a^3_1, \qquad A^3 = a^2_1 - a^1_2; \tag{2}$$

and the first field-component of $Curl^2 \vec{f}$ is

$$(Curl^2 \vec{f})^1 = A^3_2 - A^2_3$$

$$= a^2_{12} - a^1_{22} - a^1_{33} + a^3_{13} . \tag{3}$$

For the right side we have

$$(\text{Grad Div } \vec{f})^1 = a^1_{11} + a^2_{21} + a^3_{31} , \qquad (4)$$

$$(\Delta\vec{f})^1 = a^1_{11} + a^1_{22} + a^1_{33} . \qquad (5)$$

By the equality of mixed partials the difference (4) less (5) is (3), QED.

Bibliography

[1] Courant, R. <u>Differential and Integral Calculus</u>, Interscience,
      New York, 1937.

[2] Fleming, W. <u>Functions of Several Variables</u>, 2nd ed., Springer Verlag,
      1977.

[3] Nickerson, H.K., Spencer, D.C., and Steenrod, N.E. <u>Advanced Calculus</u>,
      Van Nostrand, 1959.

[4] Santalo, L.A. <u>Introduction to Integral Geometry</u>, Publications de
      l'Institut Mathématique de l'Université de Nancago,
      Hermann et Cie., Paris, 1953.

[5] Schreier, O., and Sperner, E. <u>Introduction to Modern Algebra and
      Matrix Theory</u>, Chelsea, New York, 1959.

[6] Spivak, M. <u>Calculus on Manifolds</u>, Benjamin, 1965.

Universitexts

O. Endler
Valuation Theory
1972. xii, 243p.

W.R. Fuller
FORTRAN Programming
A supplement for Calculus Courses
1977. xii, 145p. 23 illus.

W. Greub
Multilinear Algebra.
Second Edition
1978. In preparation

H. Hermes
Introduction to Mathematical Logic
Translated from the German by D. Schmidt
1973. xi, 242p.

Y.-C. Lu
Singularity Theory and an Introduction
to Catastrophe Theory
1976. xii, 199p. 81 illus.

D.A. Marcus
Number Fields
1977. viii, 279p.

J.T. Oden and J.N. Reddy
Variational Methods in Theoretical Mechanics
1976. x, 304p.

J. Schnakenberg
Thermodynamic Analysis of Biological Systems
1977. viii, 143p. 13 illus.

H. Tolle
Optimization Methods
Translated from the German by W.U. Sirk
1975. xiv, 226p. 106 illus.

Functions of Several Variables
2nd Edition

by **W. Fleming**

1977. xi, 411p. 96 illus. cloth
Undergraduate Texts in Mathematics

The new edition of *Functions of Several Variables* is an extensive revision. Like the first edition, it presents a thorough introduction to differential and integral calculus, including the integration of differential forms on manifolds. A new chapter on elementary topology makes the book more complete as an advanced calculus text; sections have been added introducing physical applications in thermodynamics, fluid dynamics, and classical rigid body mechanics.

Many mathematicians were first introduced to basic analysis by the original edition. The second edition makes this material available, in improved form, to a new generation of students.

Contents

Euclidean Spaces
Elementary Topology of E^n
Differentiation of Real-Valued Functions
Vector-Valued Functions of Several Variables
Integration
Curves and Line Integrals
Exterior Algebra and Differential Calculus
Integration on Manifolds
Appendix 1: Axioms for a Vector Space
Appendix 2: Mean Value Theorem; Taylor's Theorem
Appendix 3: Review of Riemann Integration
Appendix 4: Monotone Functions